Schriften der Mathematisch-naturwissenschaftlichen Klasse
der Heidelberger Akademie der Wissenschaften
Nr. 17 (2005)

Hans Günter Dosch
Volkhard F. Müller
Norman Sieroka

Quantum Field Theory in a Semiotic Perspective

 Springer

Hans Günter Dosch

Institut für Theoretische Physik
Universität Heidelberg
Philosophenweg 16, 69120 Heidelberg, Germany
h.g.dosch@thphys.uni-heidelberg.de

Volkhard F. Müller

Fachbereich Physik
der Technischen Universität Kaiserslautern
Postfach 3049, 67653 Kaiserslautern, Germany
vfm@physik.uni-kl.de

Norman Sieroka

ETH Zürich
Professur für Philosophie
ETH Zentrum RAC G16, 8092 Zürich, Switzerland
sieroka@phil.gess.ethz.ch

Library of Congress Control Number: 2005931712

ISBN 3-540-28211-4 Springer Berlin Heidelberg New York

Springer is a part of Springer Science+Business Media
springeronline.com
© Springer-Verlag Berlin Heidelberg 2005
Printed in Germany

Cover design: Erich Kirchner, Heidelberg
Typeset in LaTeX by the authors
and edited by PublicationService Gisela Koch, Wiesenbach, using a modified
Springer LaTeX macro-package.

Printed on acid-free paper 32/3150 Ko – 5 4 3 2 1 0

Contents

1

Introduction

In this essay we discuss epistemological implications of relativistic quantum field theory. The empirical domain of such a theory is formed by phenomena ascribed to *subnuclear particles*, sometimes still called *elementary particles*. This latter more traditional designation reflects the lasting desire of physicists to eventually find and isolate irreducible constituents of matter. Going down to the atomic level, electrons appear to play such a rôle, whereas the nuclei of atoms can be considered as compound systems of protons and neutrons, i.e. of two species of particles. This view makes sense, since the respective number of these two types of constituents essentially identifies an atomic nucleus. Extracted from a nucleus, however, the 'free' neutron is an unstable particle: it decays spontaneously into a proton, an electron and an anti-neutrino. In the past fifty years or so basically the bombardment of matter by protons or by electrons in specially devised experiments has revealed a large variety of further subnuclear objects. Successive generations of accelerators and refined collision devices provided higher and higher collision energies. All these subnuclear objects are termed 'particles' in the physics community, nearly all of these objects are unstable and decay spontaneously into other ones. The respective lifetimes of the distinct types, however, differ widely, ranging from relatively long (10^3 sec) to extremely short (10^{-25} sec). Because of this huge disparity in lifetime the notion of a particle deserves particular attention, a point laid stress on in our consideration. The study of the physical behaviour of these subnuclear particles led to distinguish three types of interactions: the *strong*, the *electromagnetic* and the *weak interaction*. As the names suggest these interactions differ in their respective strength. Furthermore, each type shows characteristic conservation laws obeyed in the observed reactions of the subnuclear particles. On the theoretical side the *Standard Model of particle physics* has emerged in the course of time. This striking achievement is supposed to account for the full hierarchy of the strong, the electromagnetic and the weak interaction.[1]

---

[1] A very lucid, non-technical overview of the subnuclear world and of basic elements of its theoretical representation can be found in Veltman (2003).

Since we are solely interested in firmly based conclusions, we confine ourselves to mathematically coherent and experimentally very well corroborated aspects of quantum field theory. Therefore, we focus on various aspects related to the Standard Model of particle physics and leave out all speculations presently in vogue, as e.g. string theory – interesting as they might be. The Standard Model is considered to essentially describe the realm of subnuclear particles up to the current experimental limit energy, probing distances down to 10^{-16} cm. We are aware of some indications – both experimental and theoretical – that the Standard Model should be modified. We believe, however, that future developments which might crystallise from today's more speculative investigations, will fit neatly in the epistemic scheme we propose, too. This holds also for the variety of partial models, motivated by the Standard Model but augmented by crucial additional assumptions or approximations; these models are not considered here either.

The principal aim of this essay is to show that various facets of quantum field theory do enter into the theory of subnuclear phenomena and deal with complementing domains. In our opinion these various facets are not always adequately considered in the existing philosophical literature, where most emphasis is put on perturbation theory. In particular, the notion of a particle is highly intricate, as will be exposed; for physical reasons there is nothing like a unique definition. Even more, the relation between particles and quantum fields is an implicit one, determined by the inherent dynamical content of the theory envisaged.

Our essay is organised as follows: chapter 2 presents a concise general description of relativistic quantum field theories viewed as physical theories, avoiding technical formulations as far as possible. Various facets are elucidated and distinguished according to differing aims pursued. In chapter 3 we return in greater detail to the central question: in which way do particles, i.e. the objects observed experimentally, emerge in the theory from the quantised fields, i.e. from the theoretical building blocks? The resulting rather complex answer manifests, to which extend the theory copes with the basic phenomenon of relativistic physics, that mass can be transmuted into energy and vice versa. It is important to notice that our considerations envisage from the very beginning quantum theories with their probabilistic physical interpretation, determined by expectation values; hence, we do not discuss the so-called particle-wave dualism. In chapter 4 we trace back the epistemic discussion to the time, when the notion of the classical electromagnetic field gained an autonomous status, freed from the attempts of a mechanical foundation. There, we encounter the birth of the symbolic interpretation of physical theories – notably in the work of Heinrich Hertz. In chapter 5 we look at the various facets of quantum field theory from a semiotic perspective. In chapter 6 we give a short résumé.

2

Relativistic Quantum Field Theories
Viewed as Physical Theories

2.1 The Empirical Domain

Phenomena ascribed to subnuclear particles and their interaction form the physical domain of relativistic quantum field theory. Such a subnuclear particle is identified by its mass and spin, which determine its behaviour under space-time transformations, and by its electric charge and further charge-like inner quantum numbers. When a stable particle is isolated from external perturbations it moves freely with constant velocity: its energy E and its momentum vector p satisfy the relativistic kinematical relation

$$E^2 = m^2 c^4 + c^2 p^2, \tag{2.1}$$

where m is the rest mass of the particle and c the vacuum velocity of light. Eventually, the particle is recognised when it triggers an appropriate (macroscopic) detection device. These recordings constitute the empirical data to be met by a relativistic quantum field theory.

The salient feature that characterises the interaction of subnuclear particles is the possible transmutation of matter into energy and vice versa: particles can be created or annihilated, provided certain conservation laws are respected – as for instance charge conservation – when a mass m is converted into its energy equivalent $\Delta E = mc^2$ or vice versa. Hence, definite configurations of particles exist only asymptotically in time before and after a process of collision, when the particles are well separated and mutually non-interacting, due to the short range of their interaction. Because of these properties, scattering experiments, where a certain initial state of particles is carefully prepared and the resulting final states are analysed by measurements, play such an important rôle in particle physics.

Many of the objects, however, which are conveniently called subnuclear particles, are not stable but decay spontaneously into lighter particles. Thus, strictly speaking, they cannot appear in an asymptotic state. Nevertheless, it is theoretically appealing and proves to be empirically justified to treat an unstable particle also as forming asymptotic states, provided its lifetime is large compared with the reaction time in a scattering process. This indicates already, that the notion of a subnuclear particle is strongly based on the related theoretical perspective. We shall be confronted with this problem several times in this essay.

2.2 The Notion of a Relativistic Quantum Field

The basic concept of the theory are quantum fields defined on space-time, not particles. Space-time is assumed to be a four-dimensional real vector space with given metrical properties and Einstein causality, such that the Poincaré group (constituted by translations and Lorentz-transformations) is implied as symmetry group. This space-time structure fixed in advance – called Minkowski space – forms the register for recording physical events. The predictions of a relativistic quantum field theory on the outcome of scattering processes are of probabilistic nature, in this respect similar to those of (non-relativistic) quantum mechanics. However, a novel feature occurs: in these processes particles can be created and annihilated. The quantum fields, in terms of which the theory is constructed, are operators that depend on space-time and act on the space of physical state vectors. This dependence on space-time, however, shows the behaviour of a generalised function or distribution. Phrased technically, a relativistic quantum field is an operator-valued distribution. Therefore, a well-defined operator cannot be related to a definite space-time point x, but only to a space-time domain of finite extent: a proper operator emerges from the quantum field $\phi(x)$ by 'smearing' with a corresponding testfunction $f(x)$:[1]

$$\phi(f) := \int d^4x \, f(x)\phi(x). \qquad (2.2)$$

The very fact, that a quantum field behaves like a distribution, has far-reaching consequences: in general, pointwise products of distributions are not defined mathematically. Hence, the nth power $\phi^n(x)$, $n = 2, 3, 4, \cdots$ of a quantum field $\phi(x)$ at a given space-time point x is a priori not defined. In contradistinction, however, such products are well defined in the realm of classical field theory, since a classical field $\varphi(x)$ is a (generally differentiable) function. Converting indiscriminately such products of classical fields into the corresponding products of quantum fields produces mathematically ill-defined

---

[1] The right hand side of the equation is only a formal representation.

objects, from which inevitably infinities emerge when used in the context of a quantum field theory. Seen historically, quantum field theory has grown out of classical field theory by the application of heuristic 'quantisation rules', extending the profitable rules that converted classical mechanics into quantum mechanics. Moreover, in order to incorporate the causal space-time structure together with the Poincaré symmetry into the theory, a Lagrangian formulation of the classical field theory is chosen. The incorporation is achieved by keeping the functional form of the classical Lagrangian density – a *local* polynomial of the field(s) and of its space-time derivatives – and just replacing the classical field $\varphi(x)$ by the quantum field $\phi(x)$. As an example we show the Lagrangian density of a model involving only a single real scalar field:

$$\mathscr{L}(x) = \frac{1}{2}\left(\partial_\mu \phi(x)\,\partial^\mu \phi(x) + m^2 \phi^2(x)\right) + \frac{g}{4!}\phi^4(x), \qquad (2.3)$$

where the first part describes a free field and the term involving the coupling constant g the (self)interaction of the field. As explained above, the ill-defined local products of field operators appearing lead to infinities, if one attempts to deduce physical consequences from such a 'theory'. In the early period of exploring quantum field theories these infinities caused a considerable amount of bewilderment. Later, a systematic reformulation of the originally ill-defined approach to relativistic quantum field theory was achieved – called perturbative renormalisation theory – which provides a mathematically well-defined quantum field theory as a formal power series in a renormalised version g_{ren} of the coupling constant. This theory produces strictly finite physical predictions in each order of the expansion. The price to be paid will be expounded in the sequel.

2.3 Local Gauge Symmetry

In classical electrodynamics, time evolution and the force acting on charged particles are directly formulated in terms of the electromagnetic field. This field can be substituted by an electromagnetic potential acting as an ancillary mathematical object. Two such potentials, however, the difference of which is a local, i.e. space-time dependent gauge transformation, provide the same electromagnetic field, and hence generate identical observable consequences. In other words, the relation between electromagnetic fields and potentials is not one-to-one: a whole equivalence class of the former corresponds to a given field, and vice versa. None the less, it is the electromagnetic potential, not the field, that enters into the Hamiltonian (and the Lagrangian) form of the classical equations of motion for charged particles. The heuristic correspondence principle – forming a 'quantisation rule' – converts this 'minimal coupling' of a particle to the electromagnetic potential into its descendent appearing in quantum mechanics. By a similar metamorphosis, the minimal coupling of a charged matter field to the electromagnetic (vector) potential in classical field theory is finally elevated to the corresponding shape in a relativistic quantum field theory. The resulting theory is quantum electrodynamics (QED). One should notice that such a heuristic theory generating process is neither tautological nor necessarily successful: the arising new theory has to turn out being physically relevant. QED stands this requirement beyond doubt. On account of its form QED is a *quantum gauge field theory*, showing a local (Abelian) gauge symmetry. This symmetry signals three basic properties of such a theory: i) only equivalence classes of the vector field are physically effective: all observables predicted by the theory, i.e all quantities having a direct physical meaning, are invariant under local gauge transformations, ii) the interaction encoded in the theory is *local*, i.e. there is no action-at-a-distance, and all physical effects propagate with finite velocity ('Nahwirkungsprinzip'), iii) performing successively two Abelian gauge transformations, they commute, i.e. the result does not depend on the order applied.

Notably, the Standard Model of particle physics is in its entirety a quantum gauge field theory: the respective interaction of the basic matter fields in each of its three sectors is generated by a corresponding gauge vector field: the strong interaction by a (non-Abelian) $SU(3)$-vector field, and the electroweak interactions by $SU(2) \times U(1)$-vector fields, which, however, because of spontaneous symmetry breaking only produce an (Abelian) $U(1)$-symmetry in the physical state space. This means, that there are three families of space-time dependent inner field transformations, which do no alter the Lagrangian of the Standard Model.

2.4 Particles versus Quantum Fields

The theoretical description of a relativistic particle in the frame of a quantum theory was introduced long ago by Wigner [2] and does not rely on the notion of a quantum field. It is rather based on the behaviour of a particle under space-time transformations, attributing to it a particular unitary irreducible

---

[2] Wigner (1939).

representation of the Poincaré group connected with mass and spin. This formulation is also used in the physical state space of a quantum field theory. To gain clarity on the rôle played by quantum fields it is important to realise that the connection between the number of fields entering into a quantum field theory and the number of particles described by it is in general very indirect. A one-to-one correspondence between fields and particles appears only in a free field theory or when a model is ab initio formulated perturbatively as a formal power series – apart from ghost(!) fields to which we come back later. In addition to its dependence on space-time and the associated symmetry (Poincaré transformations), a quantum field in general carries a characteristic set of charge-like (additive) quantum numbers, connected with 'inner' symmetries. If the Lagrangian of a quantum field theory shows such a symmetry, the related quantum number provides a conservation law imposed on the reactions between particles, as e.g. the conservation of electric charge.

In the Standard Model the weak and electromagnetic interaction are treated perturbatively. Hence, as stated before, the corresponding gauge and matter fields can be directly related to specific particles, which then are identified via their decay products. The strong interaction, however, is in general not accessible by perturbation theory. It is supposed that neither the matter fields ('quark' fields) nor the $SU(3)$-gauge fields ('gluon' fields) entering the strong interaction part of the Lagrangian, correspond to asymptotic particles – even in the restricted sense qualified before. Instead, the asymptotic particles subject to the strong interaction are supposed to be related to particular 'products' of these fields. This theoretical picture is called *confinement* and has not yet been deduced rigorously. But perturbation theory can be used under very restricted kinematical conditions, where only short distances become relevant. The empirical success of these calculations, together with model calculations, support the picture drawn for the strongly interacting particles like proton and neutron. In chapter 3 we shall further elaborate on this point.

2.5 The Facets of Quantum Field Theory

In the course of its evolution quantum field theory expanded in several directions according to various aspects focussed on: the *general theory of quantum fields* and *local quantum physics* consider the basic conceptual frame and the general properties (consequences) implied. The construction of concrete models is the subject of *perturbative renormalisation theory*, of *constructive quantum field theory* and of *lattice gauge theory*. We shall now outline these facets in turn.

2.5.1 General Theory of Quantum Fields

Parallel to the advent of rigorous perturbative renormalisation theory in the fifties and sixties, and somehow animated by the clarification achieved there, vigorous efforts have been made to formulate beyond perturbation theory a mathematically sound general framework of a relativistic quantum field theory and to analyse it rigorously. The concept of a local quantum field is preserved as the basic mathematical object, however any recourse to classical field theory is avoided. The framework is formed by few precisely defined postulates – usually called *Wightman axioms* – which give full attention to the quantum field being an operator-valued distribution. Within this rather general framework a number of physically important structural consequences have been rigorously deduced.[3] In these derivations the locality property of a field operator plays a major rôle. Among these consequences we mention the CPT-symmetry, i.e. the combined application of time reflection T, particle-antiparticle conjugation C, and space reflection P, which is always a symmetry of a quantum field theory. This occurs irrespectively of whether an individual symmetry of this triple holds or is violated. Due to it a particle and its antiparticle have equal mass, an experimentally extremely well tested prediction. Another result is the theorem on the connection of spin with statistics – this connection, describing a characteristic property of microsystems, has to be postulated separately within the realm of quantum mechanics. Furthermore, the Wightman axioms have proven sufficient to form the basis for a collision theory of the particles accounted for by a quantum field theory. (The reader is reminded that the physical domain of a quantum field theory is formed by collision processes of subnuclear particles.) Theoretically, the Wightman axioms have far-reaching mathematical implications for the correlation functions of the theory: the *Wightman functions*. These functions depend by their definition on real space-time variables, however turn out to allow an analytic continuation in the sense of analytic function theory to complex values of the space-time variables. Dispersion relations for scattering amplitudes follow from the domain attained by this continuation. Moreover, this domain contains the continuation of real time values to purely imaginary time values, thus converting the original Minkowski space into a four-dimensional Euclidean space. Thereby, from the Wightman functions result the *Schwinger functions* of the *Euclidean formulation* of the original quantum field theory. Conversely, from this Euclidean formulation, the (physical) quantum field

---

[3] Streater and Wightman (1980), Jost (1965).

theory on Minkowski space can be reconstructed (Osterwalder-Schrader Theorem[4]). Seen as a mathematical construction, the Euclidean formulation is based on functional integration. The Schwinger functions corresponding to bosonic quantum fields emerge as correlation functions of a probability measure defined on a function space of generalised functions (distributions). A related well-established Gaussian measure corresponds to a free quantum field theory and thus the correlation functions are again explicitly given. In the case of a quantum field theory showing interaction, however, the measure has to be mathematically constructed, substantiating heuristic transformation rules with roots in classical field theory. In contrast, the Euclidean correlation functions corresponding to a fermionic quantum field theory emerge from 'Berezin'-integration,[5] defined on a Grassmann-algebra. As the Euclidean formulation allows to apply powerful mathematical techniques, it strongly stimulated the study of concrete models of quantum field theory.

It goes without saying that the theory of free quantum fields satisfies the Wightman axioms. It is a widespread belief that obtaining a concrete quantum field theory beyond perturbation theory would amount to achieve a (finite) rigorous mathematical construction showing the properties required by the Wightman axioms [6] or by its Euclidean equivalent.

There is a vast variety of nonperturbative phenomenological models, which involve elements of the Wightman axioms combined with additional specific assumptions to approximately deal with selected particular phenomena observed in particle physics.

2.5.2 Perturbative Renormalisation Theory

This approach is based on the well-established theory of quantum fields without interaction – the 'free theory' – and treats the physically desired interaction as a 'small perturbation'. The classical Lagrangian, acting as precursor in the construction of a quantum field theory, contains real parameters as coefficients of the field products. In the example presented in eq.(2.3) these are denoted by m and g. As a first step towards a finite theory, the naive original substitution of the classical field φ by a quantum field ϕ is corrected by replacing ϕ with a *renormalised* quantum field according to $\phi = Z^{1/2}\phi_{\text{ren}}$

---

[4] Osterwalder and Schrader (1973), Osterwalder and Schrader (1975).

[5] Berezin (1966).

[6] More precisely: these have to be slightly extended in the case of a gauge theory, however, by introducing a Gupta-Bleuler indefinite metric formulation, in order to account for the local gauge symmetry, see e.g. Itzykson and Zuber (1980).

and, moreover, regarding m, g, Z as *bare* parameters of the theory, devoid of direct physical meaning. Within the realm of quantum field theory, renormalisation theory is a method to cure the a priori ill-defined local field products by disposing of these bare parameters by way of a controlled reparametrisation procedure. The qualification 'perturbative' means that this method is formulated within a formal perturbation expansion of the interacting quantum fields. As a consequence of the reparametrisation, in each order of the formal power series expansion all observable quantities of the theory are finite and infinities never appear, even in intermediate steps. The price to be paid is the fact, that finite (new) parameters, called *renormalised parameters*, appear in the renormalised quantum field theory. However, they are undetermined by the theory and have to be fixed by comparing the theory with experiment.[7] In a somewhat metaphoric language we can say that the bare parameters, banished by the renormalisation process, have carried with them the infinities inherent in the local products of field operators. It is the central result of perturbative renormalisation theory in four space-time dimensions, that in the class of *renormalisable quantum field theories* this number of finite undetermined parameters equals the number of bare parameters.

In retrospect, the era of a systematic perturbative renormalisation theory began with Dyson's pioneer work.[8] After a considerable evolution the rigorous BPHZ-version of perturbative renormalisation theory was accomplished in the late sixties of the last century.[9] There are also technically different, however physically equivalent formulations. Moreover, an extension to cover massless fields was also achieved, emphasised by the enlarged acronym BPHZL.[10] Technically, the BPHZL-method and its relatives make essential use of the invariance of a theory under translations on space-time. In contrast, Epstein and Glaser developed an inductive construction on Minkowski space,[11] strictly based on the locality property of a field operator, which is treated ab initio properly as an operator-valued distribution. A finite perturbation expansion (as formal power series) is established directly by a method of distribution-splitting. Since this construction does not rely on translation invariance, it can be applied when on a quantum field also acts an exter-

---

[7] In the example considered: m_{ren}, g_{ren}, Z_{ren}.

[8] Dyson (1949a), Dyson (1949b).

[9] Bogoliubov and Parasiuk (1957), Hepp (1969), Zimmermann (1970). The acronym above refers to the initials of the authors quoted.

[10] Lowenstein (1976).

[11] Epstein and Glaser (1973).

nal (classical) field,[12] or in the case of a quantum field theory on curved space-time in place of Minkowski space.[13] With regard to both its technical complexity and its physical importance for the Standard Model of particle physics, perturbative renormalisation theory culminated in the demonstration by 't Hooft and Veltman that also spontaneously broken non-Abelian gauge theories are renormalisable.[14]

One notices, that the complete perturbative construction of the quantum gauge field theories could be achieved only at the expense of introducing also *unphysical* degrees of freedom. At least up to the present, these unphysical degrees of freedom seem to prove technically indispensable in order to implement the local gauge symmetry and simultaneously maintaining the full Poincaré space-time symmetry. The recourse to these additional degrees of freedom leads to an enlarged state space with indefinite metric, which embraces the space of physical states as subspace showing a positive metric. Hence, no negative probabilities emerge for physical processes; stated technically: the scattering matrix ('S-matrix') remains unitary. Regarding these unphysical degrees of freedom more closely one observes, that in the Abelian case (QED) they are precisely formed by the pure gauge degrees of freedom of the vector (potential) field. The non-Abelian $SU(2)$-theory, however, in addition makes use of auxiliary fields: a Faddeev-Popov 'ghost'- and an 'antighost'-field.[15] These are veritable scalar quantum fields – albeit with fermionic statistics, thus violating the connection of spin with statistics to hold in the case of physical degrees of freedom.

Perturbative renormalisation theory is the approach to quantum field theory which provides its most distinctive predictions. The very high degree of agreement between specific experimental results and their theoretical description due to QED made a strong empirical case for the physicists' belief in the method of perturbative renormalisation. In a wider perspective, the fact that physicists adhere to relativistic quantum fields as a useful physical concept can be attributed to this outcome. With the electroweak sector of the Standard Model perturbatively renormalised quantum field theory provides another striking instance of a physically relevant theory – although at the price of introducing the yet to be discovered 'Higgs-boson' generating spontaneous symmetry breaking.

---

[12] Dosch and Müller (1975).

[13] Brunetti and Fredenhagen (2000).

[14] 't Hooft (1971b), 't Hooft (1971a), 't Hooft and Veltman (1972).

[15] Faddeev and Slavnov (1991).

Notwithstanding all this conspicuous success, a penetrating evaluation of the theory cannot fail to notice, that it relies (heavily) on unmet theoretical issues. The theories are constructed as formal power series only, with every indication that these series are divergent. In addition, the extent, i.e. the highest order of the power series worked out, is dictated by the human limit to perform the actual calculations involved rather than by a theoretical estimate of the order needed to analyze the given experimental data. There is a widespread opinion that the series have the mathematical status of an asymptotic expansion. Whether the divergent series proves to be *Borel-summable* in concrete cases and thus yields mathematically the exact result is important and yet unanswered. Nevertheless, even an obstinate sceptic among the physicists would in full view of these intriguing questions hesitate to downgrade perturbative quantum field theory to the level of a mere set of very efficient rules.

2.5.3 Constructive Quantum Field Theory

This approach was prompted by the mathematically unsatisfactory situation that the formal renormalised perturbation series in general do not appear to be summable, i.e. do not represent a well-defined theory. Therefore, the rigorous construction of a quantum field theory showing interaction was still an open problem. It was via functional integration in the Euclidean formulation, that these shortcomings could be surmounted to a notable extent. By the development of elaborate convergent cluster expansion methods the rigorous construction of various bosonic quantum field theories in two and three space-time dimensions has been achieved, demonstrating that the concept of a quantum field theory allows a mathematical realisation outside free fields – albeit not yet on the physical (four-dimensional) space-time.[16] Seen technically, these theories are super-renormalisable, i.e. are less singular than just renormalisable ones, but still require renormalisation, achieved nonperturbatively within their rigorous construction.

Moreover, by shaping Wilson's intuitive ideas of the renormalisation flow[17] into an inductive mathematical method, a particular fermionic quantum field theory in two space-time dimensions: the Gross-Neveu model, was rigorously constructed.[18] This model has two distinguished features: it is a

---

[16] See Glimm and Jaffe (1987).

[17] Wilson and Kogut (1974).

[18] Gawedzki and Kupiainen (1985), Feldman *et al.* (1986), Disertori and Rivasseau (2000).

genuine renormalisable theory (i.e. not being super-renormalisable) and it is ultraviolet asymptotically free.[19] Due to these properties the model has some similarity with a non-Abelian gauge theory in four space-time dimensions, the latter playing a central physical rôle. This similarity probably elevates the Gross-Neveu model above the status of a mere toy model. It is worth pointing out that the rigorously constructed correlation functions of the Gross-Neveu model are also shown to be equal to their Borel-summed renormalised perturbation series. This means that the theory is mathematically determined by its perturbation expansion.

2.5.4 Lattice Gauge Theory

A distinctive method to construct a quantum gauge field theory is lattice gauge theory, initiated by Wilson.[20] It is based on the Euclidean functional integral approach to construct a quantum field theory and *does not* resort to perturbation theory, i.e. keeps away from a formal power series expansion. This method starts with replacing the Euclidean space-time continuum by a four-dimensional lattice of points with spacing a (a hypercubic lattice) and discretises the theory in a theoretically appealing form. Basically, the matter fields are attached to the sites, whereas the vector gauge fields are connected with the links between nearest neighbour sites of the lattice. This set-up is built out of extended, i.e. not point-like, gauge-invariant objects. Its distinctive feature is to fully preserve the local gauge symmetry and, moreover, to avoid the introduction of unphysical degrees of freedom, like ghost fields. The Euclidean correlation functions of the lattice theory are mathematically well-defined. The problem thus posed is to achieve a physically sensible continuum limit of them, decreasing the lattice spacing a to zero. Although analytically inaccessible, too, the lattice theory can be studied by numerical simulation methods. These simulations form essentially an exact algorithm on finite lattices, the envisaged physical continuum values, however, can only be reached via an extrapolation.

Lattice gauge theory is presently the only approach to explore constructively a physically relevant gauge theory with strong coupling. The strong interaction sector of the Standard model, called 'nonperturbative quantum chromodynamics (QCD)', is such an instance. There, the foremost problem is to deduce the variety of observed asymptotic particles (mesons and baryons) from the basic degrees of freedom of the theory, that is from the matter fields

---

[19] See section 2.6.
[20] Wilson (1974).

(quarks) and vector gauge fields (gluons) which lack an asymptotic particle content. Already numerous physical quantities – including some particle masses – have been computed using the lattice theory.[21] The extrapolation involved to vanishing lattice spacing a is based on the observation, that *dimensionless quantities* computed with sufficiently small a become less and less dependent on a, suggesting a limit to exist at vanishing a. The results obtained up to now are rather promising and appear to confirm that quantum chromodynamics is an adequate theory of strong interaction phenomena.

2.5.5 Local Quantum Physics

This relatively recent name replaces the former 'algebraic quantum field theory', being more in accord with the theoretical approach denoted. The goal of local quantum physics is to develop a coherent conceptual frame and a corresponding consistent mathematical structure serving as the foundational basis of quantum field theory.[22] The basic concept of local quantum physics is the net of local observables, i.e. theoretical constructs that are directly related to measurements. In more detail: to every bounded space-time region is attributed an algebra of operators describing the physical observables of the region. The particular physical system manifests itself in the particular relations to hold between the algebras of different regions. The salient feature of this approach is the principle of locality: two operators representing observables of space-like separated regions commute. This principle is sometimes called 'Einstein causality', as it also restricts the propagation of quantum effects not to exceed the speed of light. Quantum fields have no basic status in local quantum physics, since they do not directly represent physically observable quantities. They act only as particular building blocks of observables. In this sense, their rôle has some similarity with coordinates in describing a geometrical configuration.

Among the principal achievements of local quantum physics one finds the theory of superselection structure inherent in relativistic quantum theories and a general property (KMS-condition) of such theories in the case of finite temperature. As already expounded, all three sectors of the Standard Model are invariant under separate respective local gauge transformations of the quantum fields that compose this model. By definition, observables are invariant under these transformations. Hence, in a theory strictly based on the concept of observables such transformations remain off stage. In view of

---

[21] See Montvay and Münster (1994).

[22] A recent review can be found in Buchholz and Haag (2000).

the physical pertinence of the principle of local gauge invariance, the algebraic scheme has probably to be augmented in order to incorporate this vital principle – a task not yet fully accomplished.

2.6 Renormalisability versus Effective Field Theories

The singular properties of a relativistic quantum field emerge from assuming space-time to be continuous and requiring Einstein causality to hold. These properties prohibit to form local products of quantum fields directly, but require a modified construction. There are various theoretical arguments that at very small distances of the order of the Planck length $\lambda_P \approx 10^{-33}$ cm our mathematical model of space-time as a continuous classical manifold (Minkowski space) is untenable. It does not seem overly rash to doubt that such small distances can ever be probed by a direct experiment. Observing that present day experimental particle physics reaches $\approx 10^{-16}$ cm and is fully compatible with the Minkowski space as theoretical space-time structure, to maintain the latter on all scales appears to be a useful idealisation. Idealising extrapolations are a constitutive element of any physical theory: the classical field theories of elasticity and of a fluid, e.g., are successful physical theories in their respective domains of validity, both characterisable by a length of about 10^{-1} cm and larger. They ignore completely the atomic structure of matter with its characteristic length of 10^{-8} cm.

Adhering to local quantum field theories, a particular procedure – called renormalisation – is employed to define in a mathematically controlled manner powers of the singular quantum fields. Within this realm, renormalisable quantum field theories are distinguished by the property that this renormalisation procedure introduces only a finite number of free parameters into the theory, which then have to be determined from outside, e.g. by comparison with measurements. Moreover, these theories are internally consistent from largest down to smallest distances. Although the less restrictive concept of an *effective quantum field theory* could suffice in principle to meet the demands of empirical adequacy, it is a fact that the traditional preference for the criterion of renormalisability has created the Standard Model, considered to be the fundamental theory of present particle physics. It is worth pointing out that in particular the required renormalisability of the Standard Model's electroweak sector provides a stringent condition on the field content of this theory. The condition originates in the peculiarity – traditionally called the *anomaly phenomenon* – that a symmetry shown by a given classical precursor theory can be inescapably violated by the quantisation procedure. To maintain in the above-mentioned theory local gauge symmetry in conjunction with renormalisability demands the absence of anomalies. This is achieved, if the number of quark fields matches the number of lepton fields – a prediction well in accord with present day experiments.

Within the family of renormalisable quantum field theories a highly selective criterion is the property that such a theory is ultraviolet 'asymptotically free'.[23] Given this property the renormalised coupling of the theory becomes weaker and weaker with growing energy. There are strong arguments that a renormalisable quantum field theory can be well-defined (outside formal perturbation theory) only, if it is asymptotically free. In four-dimensional space-time, solely non-Abelian gauge theories show this distinguished property.[24] Although a strictly nonperturbative construction of a renormalisable relativistic quantum field theory in four space-time dimensions could not be achieved up to now, some insight into this challenging problem can be gained by inspecting the formal power series generated by the perturbation expansion as a whole. These respective series are not convergent but diverge strongly: at large order n their coefficients grow proportional to $n!$, indeed. In the case of an ultraviolet asymptotically free theory, however, these coefficients appear with an alternating sign $(-1)^n$, whereas all coefficients have the same sign, if the theory is not asymptotically free. As a consequence, the divergent expansion of an asymptotically free theory might be Borel summable and thereupon would in the end generate the theory as a well-defined mathematical structure. In contrast, if (ultraviolet) asymptotic freedom is absent, the mathematical status of the formal power series remains obscure. It would be fallacious to conclude that quantum field theories not involving non-Abelian gauge fields – QED is the foremost example – are physically meaningless. Although doubtful to exist as separate theories on a nonperturbative level, their perturbative construction might nevertheless prove physically sensible. As regards QED, its empirical success is incontestable.

To broaden the scope of quantum field theory, recently effective quantum field theories have received growing interest. One may roughly distinguish two approaches: a systematic one emerging from an internally consistent quantum field theory, and a pragmatic, exploratory one, conceiving new theories with inherent restrictions. We first describe the former approach. Given a perturbatively renormalised field theory which describes the interaction of

---

[23] Gross and Wilczek (1973), Politzer (1973).

[24] Provided the number of fermionic matter fields entering the theory does not exceed a certain limit.

several particle species having different masses, it is quite intuitive that in the low energy domain of the 'light' particles their interaction with the heavy ones does not play an important rôle. A quantitative account thereof furnishes a decoupling theorem.[25] Employing Wilson's renormalisation group approach it could be substantially extended.[26] We describe these results in some detail because of their paradigmatic rôle. They are derived in the realm of renormalised perturbation theory to any order (of the formal power series). Let the full theory involve two interacting fields with a light mass (m) and a heavy one (M), respectively. One compares the correlation functions of the light particles, derived from the full theory, with the corresponding correlation functions, derived from a truncated version of the theory, where the field of mass M has been deleted. In the simple form of the decoupling theorem the difference of two such corresponding correlation functions has at low energies the magnitude $(m/M)^2(\log(M/m))^\nu$, where ν is a natural number. However, an extended renormalisable quantum field theory with the field of mass m alone can be constructed such that its outcome for the low energy processes of the light particles only differs from the corresponding results following from the full theory in magnitude $(m/M)^{2n}(\log(M/m))^\nu$, where n is any natural number. To this end one has to take into account not only the correlation functions of the field of mass m already considered, but also correlation functions of this field with insertions of 'irrelevant' local operators formed with this field. The correlation functions of the refined theory are then given as linear combinations (depending on the desired degree n of accuracy) of correlation functions with appropriately chosen operator insertions. In this way, the presence of the field with the heavy mass M in the full theory is accounted for in the theory involving only the field with the light mass m alone, up to the accuracy stated. One should notice that the irrelevant local terms are treated as insertions and are not introduced into the Lagrangian, thus keeping the refined theory renormalisable. QED can be seen in this perspective as a subtheory of the Standard Model's electromagnetic sector.

In the second approach, the pragmatic endeavour, effective quantum field theories are intentionally devised towards a more or less narrow empirical domain. These theories in general cannot be deduced in a strict sense from an embracing more fundamental theory. They are not intended as fully fledged quantum field theories, but to account for certain low energy phenomena. Hence, they can tolerate a persistent ultraviolet cutoff or non-renormalisable

---

[25] Appelquist and Carrazone (1975).
[26] Kim (1995).

interaction terms. Fermi's original theory of nuclear β-decay and its later extension to the V-A theory of weak interactions,[27] or Weinberg's phenomenological approach to chiral dynamics [28] are typical examples of such theories. For more recent developments in a completely different context see, e.g. Intriligator and Seiberg (1996).

---

[27] Feynman and Gell-Mann (1958).
[28] Weinberg (1979).

3

Particles and Fields

The objects of particle physics are subnuclear particles, their properties and their interactions. All experimental information we have on these particles comes from scattering experiments (in the widest sense). As outlined in the previous chapter quantum field theory is *the* adequate theory to describe these scattering processes. Any epistemic interpretation of quantum field theory is therefore confronted with the problem of relating the field with the particle concept.[1] In the following case study we want to show how delicate this relation is, here we shall only use internal, that is physical criteria, but we shall show in the following chapters that semiotic concepts are useful for a meaningful interpretation, since the definition of a subnuclear particle is by no means obvious; we first state different possible approaches.[2]

- If one insists that there is a definite relation between energy and momentum, as in Wigner's general definition of a particle, the invariant mass of the particle must have a definite value, see eq. (1). This in turn implies that the particle must be stable. Only in that case there exists a definite invariant mass. Unstable particles show a certain spread in the mass, the so called mass width. Under the severe restriction of stability only the electron, the proton[3] together with the respective antiparticles, and the photon qualify as subnuclear particles, since they are stable to our knowledge.[4]

- Since particles are the fundamental objects as far as experiments are concerned, a practical definition might be based on the requirement that the

---

[1] As mentioned in the Introduction we do not have in mind here the so called 'particle-wave' dualism encountered in quantum mechanics.

[2] For somewhat different approaches see Cao (1996).

[3] A possible finite lifetime longer than the age of the universe does not need to concern us here.

[4] We do not dwell on the so called infrared problems, which are related to the interaction of charged particles with very low energetic photons, since these problems are solved in a very compelling model (Bloch-Nordsiek).

lifetime of a particle is long enough to construct beams of these entities and to experiment with them. In this way besides electrons, protons and photons also the π-mesons and K-mesons and hyperons e.g. qualify as particles.

– A less stringent definition only demands a 'rather sharp distribution of the rest mass' or – expressed more formally – a pole appearing in a corresponding scattering amplitude not too far from the real axis. This was a point of view much advocated in the sixties and explains the entry 'Particle listings' in the very important annual *Review of Particle Physics*.[5] This definition has a rather fuzzy frontier as becomes already clear from the appearance and disappearance of 'particles' in this listing.

– A definition of a particle which is better suited for the theoretical framework is the following: We call a particle an object which were stable if weak and electromagnetic interactions were absent. In this way also the intermediate bosons with a width of around 2.5 GeV, corresponding to lifetimes of around $3 \cdot 10^{-25}$ seconds, qualify as particles, whereas the heavy J/ψ-meson with a width of 0.000084 GeV, corresponding to the much larger lifetime of $7.5 \cdot 10^{-21}$ seconds, does not.

This shows that questions like 'What is a particle?' or 'Does such-and-such particle exist?' can only be answered in a given context of theoretical concepts and experimental feasibility. We will provide further interpretation of this feature in the essay's last chapter.

Moreover, the notions of fundamental and composite particles can be introduced only in a given theoretical context. Rutherford proposed in 1920 to search for a tight bound state of a proton and an electron which would be electrically neutral. The search was successful, the neutron was discovered 1932 by Chadwick, but nobody would nowadays consider the neutron as a bound state of an electron and a proton, though it decays into a proton, an electron and a light neutral particle, the antineutrino. Also the Z^0 boson is not considered as a bound state of an electron and a positron, though it decays, among others, into these particles. On the other hand, scientists call the J/ψ-meson a bound state of a quark and an antiquark (the so-called charmed quarks), though it cannot be separated into these objects. The question of compositeness is in our present understanding very closely connected with the relation between the concepts 'particle' and 'quantised field'.

---

[5] Eidelman and others (2004).

As mentioned in the preceding chapter, the theoretical framework of particle physics is quantum field theory and we have to interpret the relation of the basic theoretical concepts, the quantum fields, and the entities with which experiments are done, that is particles. This relation is only explicit in the model of a free field theory, that is a field theory without any interactions, see section 2.4. In a free field theory the field operator is directly constructed in terms of creation and annihilation operators of a given particle-antiparticle species. In view of the rather simple renormalisation rules for the product of two field operators in a free theory this means that in practice the particle properties can be read off from the classical Lagrangian. Free theories are unrealistic, however, and treating interaction by perturbation theory is by far the most developed part of relativistic quantum field theory. Here again we can first consider the so called tree approximation, where the simple renormalisation procedures of the free theory are applied. It corresponds to the lowest order in the perturbative expansion. In this approximation a unique identification of particles with fields is also possible. For the leptonic part of the Standard Model of electroweak interactions the quantum fields occurring in the free part of the Lagrangian correspond to the 'observed' particles. Whereas for the gauge bosons Z^0 and $W^\pm$ the extended definition allowing for a certain energy width has to be applied in the particle definition. The fact that already the tree approximation yields good quantitative results for the width and that the higher order corrections using renormalised perturbation theory give excellent agreement with experiment shows the fruitfulness of that approach.

The situation is much less clear for the strong interaction. Progress in the last 30 years resulted from abandoning the concept that the observed particles are directly related to quantum fields occurring in the theory. Rather it is hoped that the properties of the observed particles (hadrons) can be calculated within a theory the basic fields of which do not correspond to particles. These basic fields are the quark and gluon fields. This new concept has roots in four different facets of quantum chromodynamics and can lead to quite different concepts of 'elementary particles':

- 1) **Perturbation theory:** In certain reactions and under certain kinematical conditions a hadronic amplitude can be separated (factorised) into a long range part which is not accessible to perturbation theory and a short range part which is. In that way, for instance, relations between scattering amplitudes of particles at different kinematical conditions can be calculated and compared with experiments. This has been done very successfully in deep inelastic scattering experiments. Highly energetic electrons or positrons are scattered off protons and the calculated dependence of the cross sections on (high) momentum transfer is compared with experiment. Another example are the so called jet events, where the topology of the produced observed particles reflects that of the quarks and gluons; again a quantitative comparison with experiment is possible here. The generally accepted 'proof for the existence of a gluon' by observing three jets in the annihilation of an electron and a positron with very high energies is based on such calculations. It should be emphasised that the sophisticated framework of renormalised perturbation theory is essential for such a simple assertion that the structure of three particle jets reflects the creation of a 'quark, an antiquark and a gluon'. Such three jet events formed by hadrons were observed for some time, but could not safely be related to an elementary process involving three fundamental fields. In perturbative quantum chromodynamics all amplitudes are calculated on the basis as if the quarks and gluons were particles. It is, however, noted that, in contradistinction to the electroweak theory, the perturbatively calculated strong scattering amplitude is the more affected by uncontrollable nonperturbative terms, for instance by power corrections, the nearer one comes to the singularities induced by the quark propagators. It is nevertheless possible – but by no means compelling – to extend the concept of a particle. In perturbation theory there is a well defined reciprocal correspondence between particles and fields as mentioned above. If perturbation theory can be applied to certain reactions one could say that the existence of the particles corresponding to the fields manifests itself there. We should, however, emphasise that this definition has no consequence whatsoever. The statement quarks and gluons are 'particles' is synonymous with the following: the perturbatively renormalised QCD Lagrangian containing quark and gluon fields leads to certain consequences which can be compared with experiment. It should be noted that, according to this definition, also the so called ghost fields would be promoted to describe ghost particles, since their occurrence in covariant renormalised quantum field theory is necessary.

- **2) Lattice gauge theory:** There is up to now no analytic way to treat a realistic quantum field theory like the Standard Model in a nonperturbative way. The only attempt in this direction is lattice gauge theory, see section 2.5.4. Here, observed hadron masses can be obtained indeed from numerical calculations based directly on the QCD Lagrangian, but only in the case of finite lattice spacings. The limit to the continuum, that is to full Lorentz invariance, has not yet been established mathematically. The numerical results are very encouraging, however, and many conjectures made in nonperturbative quantum chromodynamics can be checked quantitatively. In this approach the quarks and gluons enter essentially as fields but have no particle properties whatsoever,[6] since the propagators of single quarks and gluons vanish due to gauge invariance.

- **3) Non-relativistic quantum field theory:** After the discovery of heavy mesons like the J/ψ and Υ it became clear that the hypothesis that these mesons are composed of two quarks leads to similar precise predictions as in atomic physics. One treats the quarks within the nonrelativistic Schrödinger equation like particles bound by a potential and can calculate corrections to the nonrelativistic results as power series in $\frac{\Lambda_{QCD}}{M}$ where Λ_{QCD} is some typical hadronic scale and M the mass of the heavy quark. It should be noted that simple model calculations are of typical back-of-an-envelope type, but that for reliable predictions the full machinery of effective renormalised perturbation theory must be applied.

- **4) General theory of quantum fields:** From the Wightman postulates there follows no direct relation between the quantum fields entering a theory and its particle content. The latter is supposed to be a consequence of the inherent interaction and to manifest itself in particular properties of the Wightman functions.

---

[6] Even after analytic continuation due to Minkowski metric.

Theories of Signs and Symbols, and Structural Realism

The antagonism in physics between a particle as typical ingredient of a mechanistic interpretation of nature and the more formal concept of a field dates back to the middle of the 19th century. The ongoing success of Newtonian and Euler-Lagrangian mechanics led to a widely accepted mechanistic *Weltbild* during the 19th century. This *Weltbild*, however, came to a crisis through the advance of electrodynamics which later culminated in the establishment of Maxwell's equations. Ironically, the essential new ingredient and cornerstone of these equations, the displacement current, was derived in a mechanistic model[1] of the æther, the assumed carrier of electromagnetic phenomena. Therefore electrodynamics was formally embedded in the mechanistic Euler-Lagrangian field theory. It turned out, however, that a mechanistic interpretation of the æther became less and less tenable and so the concept of a field was emerging as a map of space-time points to measurable quantities without assuming a special carrier of the field. This concept existed besides the old concepts of mechanics and allowed a perfectly adequate description of electrodynamic phenomena. Since many of the epistemic problems related to quantum field theory started with the raise of electrodynamical field theory in the middle of the 19th century it is not astonishing that also many of the philosophical concepts still important to tackle modern problems came up at the same time.[2] This is particularly true for the 'structural realism' which in

---

[1] Here and in the following we mean 'model' to refer to a set of assumptions adapted to a phenomenological description rather than something strictly derived from a set of mathematical postulates. The latter we would call a theory. For some examples of theories, models (and effective field theories) in the context of quantum field theory see Hartmann (2001).

[2] Presumably this is due to the fact that along with this transition the fundamental entities involved changed quite radically (going from particles to fields) so that even basic Newtonian principles like 'action equals reaction' got lost, see e.g. Poincaré (1905), VIII, *Le principe de Newton*.

analytic philosophy is generally associated with the work of Henri Poincaré.[3] Thus, before talking about the history in more detail, let us briefly introduce the modern origin of structural realism; in particular since it is – in different ways though – invoked by several writers in the interpretation of quantum field theory at the moment.[4]

The modern origin of 'structural realism' lies in the tension between what is called the miracle argument and the pessimistic meta-induction. The miracle argument expresses the idea that scientific realism is the only philosophy of science that can explain why certain theories are so successful in their predictions whereas this fact remains a miracle for all those accounts that view theories and the entities they deal with as constructions of our mind in some sense. For why should our mental constructions depict natural phenomena so well? The pessimistic meta-induction on the other hand is the inference that, since all former scientific theories have turned out to be wrong if applied generally, our current theories will also turn out to be wrong . This, of course, is particularly worrying for the realist. Thus his concern becomes this, in the words of John Worrall : 'is it possible to have the best of both worlds, to account (no matter how tentatively) for the empirical success of theoretical science without running fool of the historical facts about theory change?'[5]

Structural realism about science is the view that most physical theories are structurally correct. This is not to say – as entity realism does – that physical theories mirror the objects of a mind-independent world directly but merely that the nature of the linkages they introduce is true. The easiest way to understand what is meant here by 'linkage' or 'structure' is to have a look at the mathematical equations that are involved. Often old theories are mathematically limiting cases of new theories – the notorious example in the literature being the transition from Fresnel to Maxwell in the description of optical phenomena. Thus, structural realism seems to be more cumulative and more promising in coping with the pessimistic meta-induction.

---

[3] Although the term 'structural realism' itself was presumably coined in papers by Grover Maxwell in 1970. However, Maxwell developed his approach by considering later works of Russell and by using Ramsey-sentences (i.e. an entity is 'whatever it is' that satisfies the relations held by predicate variables in the Ramsey-sentence).

[4] Quite straightforwardly, for instance, in Cao (2001), and more formally elaborated (along the lines of Hermann Weyl to which we will also come back later), for instance, in Auyang (1995).

[5] Worral (1996), p. 151.

A further reason why talking about structures or relations seems fundamental for physics is that our experiments find out about the way in which the things under investigation *act*, i.e. the way they impinge on our measuring devices. Hence, one has to acknowledge the possibility that we might be unable to distinguish between two quite different intrinsic properties, namely if they are exactly the same with respect to the way they impinge on our instruments. By the same token we must then accept the possibility that we know next to nothing about the intrinsic nature of physical objects. Also the fact that we normally observe coherence in properties, that certain structural properties seem to be tied together, is of little help here.[6] For the fact that some relational properties come in bundles is no guarantee for that bundle to describe the intrinsic nature of a physical entity.

In the following we give a short historical sketch of concepts which we think are not only the roots of structural realism[7] but are also the basis for a semiotic approach going further than structural realism. We start with the theory of signs (symbols) of Hermann von Helmholtz and Heinrich Hertz and then find some further development in the works of Henri Poincaré and Hermann Weyl. Obviously we do not give this historical reconstruction only for its own sake, but because we think that in comparison to structural realism it will provide us with a more appropriate perspective to evaluate quantum field theory as a physical theory. Thus, this essay's implicit critique on modern analytic structural realism is indeed a systematic rather than a historical one. Of course, we think it to be important to 'get history right', but as a critique against Worrall etc. this would grasp at thin air to some extend. For although Worrall and others mainly discuss Poincaré they admittedly did not attempt to give a historical reconstruction of the origin of structural realism.

There was a geographical distribution of the prevalence of either the mechanistic or the more formal epistemological concept of physics. The former one was especially flourishing on the British islands whereas the latter one was more popular on the continent, notably in France and Germany. This difference between the British and continental epistemological concept is extensively discussed by Duhem in his book *La Théorie Physique*.[8] Helmholtz writes that 'English physicists like Lord Kelvin and Maxwell'[9] were more satisfied by mechanical explanations than by the 'plain most general expla-

---

[6] Such an argument has been brought forward, e.g., in Chakravartty (2002).

[7] See also Cao (1996).

[8] Duhem (1914 1957), I, IV *Les théories abstraites et les modèles mécaniques*.

[9] Of course, both are Scottish.

nation of the facts and their laws as they are given in physics by the systems of differential equations.'[10] He himself, however, adhered to this formal representation and felt most assured by that.

We have mentioned the argument brought forward in favour of structural realism that we know next to nothing on the *intrinsic* nature of physical objects, but only how they act on our measuring devices. Helmholtz anticipated this argument by applying it already to our sensual receptions;[11] in that he was largely influenced by the 'Sinnesphysiologie' of Johannes Müller.[12] Helmholtz writes:

> Our sensations are actions which are evoked by external causes in our organs, and how such an action manifests itself depends of course essentially on the nature of the apparatus which is acted on. In so far as our sensation gives us a message of the peculiarity of the evoking external influence it can be accepted as a sign of it but not as a copy.[13]

The essential features of his theory of signs can be found in his 'Handbook of Physiological Optics' the most elaborate version is in his lecture 'Die Thatsachen in der Wahrnehmung' where the above quotation is taken from. Though he insists that our sensual impressions are only signs they are not to be disposed of as empty 'Schein' but they are signs of *something*, be it existing or happening, and what is the most important, they map the *law* of what is happening. For Helmholtz the relevant feature of science is not the particular set of signs, but that what he calls 'law' and which he defines as the unchanging relation between changing variables (signs). In this respect his above quoted assessment of the differential equations as the more assuring feature of a theory becomes clear.

Helmholtz did not try do develop an elaborate epistemological scheme, his ideas were rather intended to give a basis to his proper work which embraced physiology, physics and mathematics. His main concern is a warning against

---

[10] Foreword to Hertz (1894).

[11] Recently the same point has been made in Psillos (2001), p. S14.

[12] Helmholtz (1892), Helmholtz (1921).

[13] 'Unsere Empfindungen sind Wirkungen, welche durch äussere Ursachen in unseren Organen hervorgebracht werden, und wie eine solche Wirkung sich äussert, hängt natürlich ganz wesentlich von der Art des Apparates ab, auf den gewirkt wird. Insofern die Qualität unserer Empfindung uns von der Eigentümlichkeit der äusseren Einwirkung, durch welche sie erregt ist, eine Nachricht gibt, kann sie als ein Zeichen derselben gelten, aber nicht als Abbild.' Helmholtz (1892), vol II, p. 226.

drawing metaphysical conclusions from results of natural science and attaining a dogmatic stance on them. He concedes that hypotheses are necessary for scientific research and even more for human action. In this sense he regards also metaphysical hypotheses as possibly legitimate. But he warns: 'But is unworthy of a thinker who wants to be scientific if he forgets the hypothetical [...] origin.' [14]

Heinrich Hertz was a student of Helmholtz. He contributed largely to the development of electrodynamical field theory, both by his experimental and theoretical investigations. Einstein, for instance, referred to the 'Maxwell-Hertz theory' of electrodynamics in his writings. Like Helmholtz, Hertz was more satisfied by mathematical equations than by mechanistic pictures. On Maxwell's equations he writes admiringly: 'One cannot read this beautiful theory without sometimes feeling as if those mathematical formulæ had their own life and intelligence, as if those were more clever than we, even more clever than their inventor, as if they would yield more than was put into them at the time.' [15]

Hertz considers as the principal aim of conscious natural science (bewusster Naturerkenntnis) to foresee future experiences. In order to reach that he proposes a 'sign theory' which is more explicit than Helmholtz's and he gives a set of rules, both formative and descriptive.

> We form for us inner phantoms (Scheinbilder) or symbols of the things and in such a way that the logical (denknotwendig) consequences of the symbols are always pictures of the physically necessary (naturnotwendig[16]) consequences of the depicted objects.[17]

---

[14] 'Unwürdig eines wissenschaftlich sein wollenden Denkers aber ist es, wenn er den hypothetischen [...] Ursprung vergisst.' Helmholtz (1892), p. 243.

[15] 'Man kann diese wunderbare Theorie nicht studieren, ohne bisweilen die Empfindung zu haben, als wohne den mathematischen Formeln selbständiges Leben und eigener Verstand inne als seien dieselben klüger als wir, klüger sogar als ihr Erfinder, als gäben sie uns mehr heraus, als seinerzeit in sie hineingelegt wurde.' Hertz (1889), p. 11

[16] Hertz himself uses the terms 'denknotwendig' and 'naturnotwendig' instead of 'logisch' and 'gesetzlich'. However, given the context it is obvious that he does not use 'necessity' in the way we do it today when talking about laws of nature. Something similar holds for his use of 'essential' which does not mean 'metaphysically prior' in Hertz but something like 'experimentally settled'.

[17] 'Wir machen uns innere Scheinbilder oder Symbole der äusseren Gegenstände, und zwar machen wir sie von folgender Art, dass die denknotwendigen Folgen

He does not take it for granted that such a procedure is possible, but notes that experience tells us, that it is. Hertz gives several selection criteria for the symbols. Although those criteria are more important for the gathering of models than for the development of a theory (which is what we are mainly interested in), we think that they are worth mentioning. Hertz writes:

> The images we speak of are our imaginations of the things; they have with the things the one essential correspondence which lies in the fulfillment of the above mentioned postulates.[18]

He also points out that by this congruence requirement the construction of the pictures is by no means unique, but that different pictures can be distinguished according to their 'admissibility' (*Zulässigkeit*), 'correctness' (*Richtigkeit*) and 'appropriateness' (*Zweckmäßigkeit*). Pictures are admissible if they do not contradict logic (laws of thought), and admissible pictures are correct if 'their essential relations do not contradict the relations between external objects'. However, admissible and correct pictures can (and normally will) differ in their appropriateness which comes in two guises; namely as 'distinctness' (*Deutlichkeit*) and 'simplicity' (*Einfachheit*). A picture which mirrors more of an object's essential relations is called a clearer one, and amongst equally clear pictures the one with the less vacuous relations (i.e. relations which occur in the picture but not in the external world) is the simpler one.[19] Such competing pictures normally occur if one is simplifying certain aspects of a more complete theory in models. These criteria as distinctness and simplicity can be important also in more fundamental theoretical questions, for instance the relation between conventional quantum field theory and the approach by the algebra of local observables which concentrates exclusively on the local *observables*, see chapter 2.5.5.

There is a crucial difference between the sign theory of Hertz and that of Helmholtz. Helmholtz's signs are related to the sensual impressions whereas those of Hertz can be free creations of the mind. Here a semiotic distinction drawn by Charles Sanders Peirce is helpful to contrast between these two types

---

der Bilder stets wieder Bilder seien von den naturnotwendigen Folgen der abge-bildeten Gegenstände.' Hertz (1894), pp. 1f.

[18] 'Die Bilder, von welchen wir reden, sind unsere Vorstellungen von den Dingen; sie haben mit den Dingen die eine wesentliche Übereinstimmung, welche in der Erfüllung der genannten Forderung liegt.' Hertz (1894), p. 2.

[19] Note that Hertz's definition of 'simplicity' is not like the common blurry one, that one wouldn't like as a functional relation a polynomial of 27th order or something alike.

of 'signs'; namely the distinction between indexical signs and conventional signs (symbols).[20] The former show some existential relation to their objects of reference; i.e. their meaning is based on a cause and effect relationship. Thus, for instance, a weather cock is an indexical sign of the wind since it is turned by the wind into its direction. A symbol on the other hand lacks such a cause and effect relationship. Its meaning is based on a conventional regularity as, for instance, is our use of the word 'strawberries' to refer to strawberries. For obviously there is neither any causal relation nor any resemblance between the string of letters in 'strawberries' and those red and eatable objects. Also the relation between a quantised field, the essential ingredient of the theory and a particle, on which the interpretation of experiments is based on, can be very indirect, as discussed in the preceding chapter.

Thus, one can take Helmholtz's approach, which started from sense perception, to be a theory of indexical signs; whereas Hertz's theory should already count as one of symbols.

The historical development went further away from a theory of indexical signs and more and more towards a theory of symbols (conventional signs). Both Helmholtz and Hertz were certainly motivated by the unreconciled differences of classical mechanics and of Maxwell's theory but they had no reason to doubt the unconditioned physical validity of classical mechanics and Maxwell's theory. This was no longer the case for the next main contributor in this tradition, the great mathematician Henri Poincaré who lived long enough to see the beginning erosion of classical physics at the turn to the 20th century. He expounded his views on the epistemological foundation of science in three works: *Science et l'hypothèse* (1903), *La valeur de la science* (1905), and *Science et méthode* (1908). Very much like Helmholtz and Hertz he takes the success of science as a starting point and concludes – in accordance with today's miracle argument – that this success would not be possible if science would not give us knowledge of something of reality (*quelque chose de la réalité*). However, he continues in line with what is today called the pessimistic meta-induction, that what science can attain are not the things themselves, as the naive entity realists think, but only the relations between the things (*les rapports entre les choses*). Outside these relations there

---

[20] Peirce also introduces 'icons' (iconic signs) which have some relation of resemblance to their denoted object. However, since straightforward resemblances are already ruled out in Helmholtz's approach, we do not need to take icons into account here.

is no recognizable (*connaisable*) reality.[21] Thus, this strain in Poincaré's work is the starting point of nowadays structural realism as introduced above. However, this is mainly based on a single passage in which Poincaré discusses the transition from Fresnel's to Maxwell's description of optical phenomena; it is closely related to Hertz's introduction of signs mentioned above:

> [...] the aim of Fresnel was not to find out whether there really is an ether, whether it is or it is not formed of atoms, whether these atoms really move in this or that sense; his object was to foresee optical phenomena.
>
> Now, Fresnel's theory always permits of this, to-day as well as before Maxwell. The differential equations are always true [...] They teach us, now as then, that there is such and such a relation between some thing and some other thing; only this something formerly we called *motion*; we now call it *electric current*. But these appellations were only images substituted for the real objects which nature will eternally hide from us. The true relations between these real objects are the only reality we can attain to, and the only condition is that the same relations exist between these objects as between the images by which we are forced to replace them. If these relations are known to us, what matter if we deem it convenient to replace one image by another.[22]

---

[21] 'Cela ne pourrait être si elle ne nous faisait connaître quelque chose de réalité; mais ce qu'elle peut atteindre, ce ne sont pas les choses elles-mêmes, comme les pensent les dogmatistes naïfs, ce sont seulement les rapports entre les choses; en dehors de ces rapports, il n'y a pas de réalité connaisable.' Poincaré (1927), Introduction, p. 4.

[22] 'Nulle théorie semblait plus solide que celle de Fresnel qui attribuait la lumière aux mouvements de l'éther. Cependant, on lui préfère maintenant celle de Maxwell. Cela veut-il dire que l'oeuvre de Fresnel a été vaine? Non, car le but de Fresnel n'était pas de savoir s'il y a réellement un éther, s'il est ou non formeé d'atomes, si ces atomes se meuvent réellement dans tel ou tel sense; c'était de prévoir les phénomènes optiques. Or, cela, la théorie de Fresnel le permet toujours, aujourd'hui aussi bien qu'avant Maxwell. Les équations differentielles sont toujours vraies; [...] Elles nous apprennent, après comme avant, qu'il y a tel rapport entre quelque chose et quelque autre chose; seulement, ce quelque chose nous l'appelions autrefois *mouvement*, nous l'appelons maintenant *courant électrique*. Mais ces appelations n'étaient que des images substituées aux objets réels que la nature nous cachera éternellement. Les rapports véritables entre ces objets réels sont la seule réalité que nous puissions atteindre, et la seule condition, c'est qu'il y ait les mêmes rapports entre ces objets qu'entre les images que nous sommes

Obviously one can read this quote in two ways, namely as making purely epistemic claims or as making metaphysical claims (this will mainly depend on the translation and reading of 'The true relations between these real objects are the only reality we can attain to [...]'). Thus, taking only this passage into account, Poincaré's 'structural realism' could come in two versions: according to modern terminology one is called *epistemic* (or restrictive) structural realism and the other *ontic* (or eliminative) structural realism.[23] The most prominent advocates of epistemic structural realism are Worrall and Zahar, while ontic structural realism is argued for notably in the works of French and Ladyman. See, e.g., Worral (1996), Zahar (1996), and Ladyman (1998). The two views differ about the question as to whether there is something else in the world than structure. While ontic structural realism says no, epistemic structural realism allows for entities that stand in the relations we find out about. However, according to the latter we will never know about those entities; i.e. nature will eternally hide from us the *things in themselves* (cf. what we said about the detection of intrinsic properties earlier on). Thus one might call epistemic structural realism a 'Kantian physicalism'.[24]

The main critique that has been brought forward against ontic structural realism – and with which the authors concur – is that one cannot have relations without relata. Thus, we are more sympathetic to its epistemic version which is also much more in line with Poincaré. For Poincaré stresses the importance of conventions in science – indeed he is notoriously credited as the very founding father of conventionalism (and surely not of realism!). He illustrates the rôle of conventions in science by the following thought experiment. Assume a fictitious earth covered with thick clouds such that the stars were never visible. Though scientists living on that fictitious earth would finally find out that many facts can be explained by assuming that this earth turns itself around an axis, this statement is a pure convention. The statements 'the earth turns itself' and 'it is more convenient to assume that the earth turns itself' have accordingly one and the same meaning.[25] He also stresses that the choice of geometry is purely conventional. He explicitly says that the axioms of Euclid are neither synthetic judgments *a priori* nor experimental results; they are *conventions*.[26]

---

forcés de mettre à leur place. Si ces rapports nous sont connus, qu'importe si nous jugeons commode de remplacer une image par une autre.' Poincaré (1927), X, p. 190.

[23] A distinction drawn by Ladyman (1998) and by Psillos (2001).

[24] See Jackson (1998), pp. 23–24.

[25] Poincaré (1927), VIII p. 139, Poincaré (1905), XI, sect. 7, pp. 271ff.

[26] Poincaré (1927), III p. 66.

Hence, this conventional and not the realistic aspect was the dominant one in his philosophy.[27] This can also be seen from another passage in *Science et l'hypothèse* where Poincaré again refers to Fresnel's theory (this time to his laws of refraction) but now stresses their approximative character.[28]

The delicacy of the balance between conventionalism and structural realism becomes clear from the following résumé he draws near the end of *La valeur de la science*:

> In summary, the only objective reality are the relations between the things (choses) from which the universal harmony starts. Doubtless these relations, this harmony, would not be conceived outside a mind who conceives or feels them. But they are nevertheless objective [intersubjective] since they are, become, or stay common to all thinking beings.[29]

Poincaré lived until 1912 and saw the big changes occurring at the turn of the century. He seems not to have been impressed strongly by the birth of quantum physics, but the effects of electrodynamics on mechanics shook him considerably, as can be read from the last chapter of *Science et l'hypothèse*, entitled 'La fin de la matière'. He also foresaw the consequences on the best established physical theory, Newton's theory on gravity: 'If there is no more mass, what becomes of the law of Newton?', he asks.

Therefore Poincaré is aware that 'theories seem fragile' but he asserts that they do not die fully and from each of them there rests something. This something has to be looked for and sorted out, but it is this, and only this, which is the 'true reality'. This process of sorting out was particularly necessary in the further development towards the theory of (special) relativity, quantum mechanics and quantum field theory. In this transition also mathematical relations which were considered to be absolutely valid physical relations turned out to be obeyed only approximately and could be violated vehemently under certain conditions. This was of course a shock for those believing in an ontological significance of the concepts of science, but even the conventionalist

---

[27] The realistic aspects are mainly brought in to prevent conventionalism to drift into what he calls 'nominalism' – cf. Poincaré (1905), X pp. 233ff in particular.

[28] Poincaré (1927), X, p. 211.

[29] 'En résumé, la seule réalité objective, ce sont les rapports des choses d'où résulte l'harmomie universelle. Sans doutes ces rapports, cette harmonie ne sauraient être conçus en dehors d'un esprit qui les conçoit ou qui les sent. Mais ils sont néanmois objectifs parce qu'ils sont, deviendront, ou resteront communs à tous les êtres pensants.' Poincaré (1905), XI, p. 271.

(and '*structural* realist') Poincaré writes somewhat bewildered on relativistic mechanics:

> The essential attribute of matter is its mass, its inertia. The mass is that which always at any place stays constant [...] If therefore one has proven that the mass, the inertia does not truly belong to it in reality [...], that this mass, the constance *par excellence* is itself susceptible to changes, then one could say that matter does not exist. Now that is precisely what one announces.[30]

The fourth eminent figure whose interpretation of science we shall discuss briefly is the mathematician Hermann Weyl. He did not only live during the period of great changes in the first decades of the 20th century but also contributed essentially to them and was fully aware of the epistemological impact. In the preface to his articles 'What is matter' (1923) he notes: 'It is perhaps especially now to early, to speak about the essence of matter, where quantum mechanics [...] keeps her secrets still in a firmly closed shell.'[31] As is also evident from the formulation 'essence of matter' he uses, at least hypothetically, metaphysical concepts. As a consequence, in his discussion of Hertz's 'Mechanik' he emphasises the actual treatment of matter in the main part of the work, rather than the symbolic programme in the introduction. He sees Hertz as a follower of Huygens in the line of the 'substance theory' of matter but he remarks that with Hertz 'the concept of substance is by means of mathematical generalisation formalised to an abstract scheme.'[32] He admits that the final systematic form of science might be of a similar kind but he doubts 'that through the cancellation of metaphysical intuition [...] to which

---

[30] 'L'attribut essentiel de la matière, c'est sa masse, son inertie. La masse est ce qui partout et toujours demeure constant [...]. Si donc on venait à démontrer que la masse, l'inertie de la matière ne lui appartient pas en réalité [...] que cette masse, la constante par excellence, est elle-même susceptible d'altération, on pourrait bien dire que la matière n'existe pas. Or, c'est là précisément ce qu'on annonce.' Poincaré (1927), XIV, p. 282.

[31] 'Es ist vielleicht gerade heute besonders *verfrüht*, über das Wesen der Materie zu reden, wo die Quantentheorie [...] ihr Geheimnis noch immer in fest verschlossener Schale hält.' Weyl (1924).

[32] Weyl (1924), p. 17. This transition from a metaphysical to a mathematical formal point of view seems to have a long tradition. Leibniz, the opponent of Huygens in matters of atomism, writes 1694 in a letter to l'Hospital that 'his metaphysics is all mathematics, so-to-speak, or could become it', Leibniz (1849), vol II, p. 258. We shall see a similar development in the arguments of Weyl.

the concept of substance belongs, theoretical interpretation looses everything stringent.'

Weyl had generalised Riemannian geometry introducing the gauge principle by which the length of a measuring device could depend on its history in space-time. In that way he could obtain a stringent common interpretation of the two great branches of classical field theory: Einstein's theory of gravitation and Maxwell's electrodynamics. But the application of Weyl's principle to geometry contradicted experiment, as especially Einstein realised immediately. The general principle of 'gauge invariance', however, turned out to be one of the most fruitful ones in quantum field theory. London, who was the first to recognise the fruitfulness of Weyl's postulates in the then newly developed modern form of quantum mechanics (1927), commented on the long adherence of Weyl to his theory:[33] 'An uncommon clear metaphysical conviction was necessary' not to accept the experimental evidence.[34] Weyl, however, immediately took up the idea that quantum mechanics was the right application of his gauge theory, in particular after Dirac had given a relativistic description of it.[35] But now it was not the length of a measuring device which was gauged, but the very abstract phase of a wave function, in quantum field theory the wave of the charged field. As we have already mentioned in chapter 2 all three sectors of the Standard Model of particle physics are gauge theories in the sense of Weyl.

In contrast to his metaphysical inclinations of his earlier writings, Weyl refers in the philosophical publications of his later years to the symbolic form as the adequate approach to mathematics and physics. This becomes clear already from the titles of his articles; 'Science as a symbolic Construction of Man'[36] and 'On the Symbolism of Mathematics and Mathematical Physics'.[37] In the former one he shows in case studies, very much like the philosopher Ernst Cassirer, the development from substantial to symbolic forms. Here he also quotes approvingly the symbolic programme of Hertz and he comes to such definite statements as:

it is the free in symbols acting spirit which constructs himself in physics a frame to which he refers the manifold of phenomena. He

[33] London (1927).

[34] The discussion between Einstein and Weyl has been documented by Straumann (1987), see also O'Raifeartaigh and Straumann (2000).

[35] Weyl (1929).

[36] Wissenschaft als symbolische Konstruktion des Menschen. Weyl (1949).

[37] Über den Symbolismus der Mathematik und mathematischen Physik. Weyl (1953).

does not need for that imported means like space and time and particles
of substance; he takes everything from himself.[38]

He refers to the attempts of Alfred North Whitehead and Bertrand Russell
to base reality on *events* and considers them as 'Loves Labour Lost' since we
are left with nothing than symbolic construction and that the recent developments in modern physics by general relativity and quantum theory confirmed
that.[39]

One aim of our essay is to corroborate this point of view analysing especially the developments of quantum field theory in the 80ies of the 20th
century. However, since we have to talk about several symbolic representations, it will be helpful to briefly consider one further idea brought forward
by Weyl to evaluate the 'objective core' of different representations.[40] The
basic idea is that there are always (trivial) differences between different representations, but that there are also certain features which are kept stable if a
transformation from one representation to another occurs.[41] If the origin of
some geometric object represented in a Cartesian coordinate system is shifted
or the axes are rotated the absolute coordinates of the object will change, too.
However, the shape of the figure (the length of its sides etc.) will stay the
same. Thus, the length of the sides rather than a certain abscissa value is an
objective property of the figure. In Weyl's more technical terms:

> What we learn from our whole discussion and what has indeed become
> a guiding principle in modern mathematics is this lesson: *Whenever
> you have to do with a structure-endowed entity Σ try to determine
> its group of automorphisms*, the group of those element-wise trans-
> formations which leave all structural relations undisturbed. You can
> expect to gain a deep insight into the constitution of Σ in this way.[42]

---

[38] '[...] dass es der freie, in Symbolen schaffende Geist ist, der sich in der Physik ein
objektives Gerüst baut, auf das er die Mannigfaltigkeit der Phänomene ordnend
bezieht. Er bedarf dazu keiner solchen von aussen gelieferten Mittel wie Raum,
Zeit und Substanzpartikel; er nimmt alles aus sich selbst'. Weyl (1949).

[39] At least with respect to Whitehead this is no adequate critique. For his philosophy
is exactly one of (the transformation of) symbols.

[40] Indeed this approach can be traced back to Helmholtz who maintained that the
special nature of a melody is it's invariance under transpositions, which is also
an essential feature of space Helmholtz (1877) XIX, p. 596 f. It was also taken
over, e.g., by Cassirer in his paper 'The Concept of Group and the Theory of
Perception'.

[41] Cf. Weyl (1952) and Weyl (2000), Ch. 13.

[42] Weyl (1952), p. 144.

Weyl also emphasises that transformation invariance and symmetry belong together and that they are important not only for mathematics and the exact sciences but also for arts etc.[43] Indeed we regard it to be important for any symbolic system and in particular that any comparison between symbolic representations can make use of this type of transformation-invariance-approach.[44]

---

[43] Cf. Weyl (1952), p. 145, e.g.

[44] Recently Weyl's transformation-invariance-approach was put into spotlight in the context of quantum field theory by Auyang (1995). Her version roughly goes like this: While intrinsic properties or 'things in themselves' – which we may denote by x – are necessarily hidden, we can encounter their functional relations. Thus, even without any knowledge of x, we can know about $f(x) = x_f$, $g(x) = x_g$ etc. If on top of that we also know the rules of transformation to get from x_f to x_g (i.e. we know the function $g \circ f^{-1}$), we will know the invariant properties under these transformations and hence the most likely candidates for objective properties (i.e. for the 'real' or intrinsic properties of x).

5

A Theory of Symbols
for Quantum Field Theory

5.1 Symbols and Relations

The aim of this chapter is to widen the symbolic interpretation of science initiated by Helmholtz, Hertz and Poincaré. We introduce some concepts from modern semiotics and apply them to quantum field theory. This will lead to a discussion on two levels of symbolic representation; namely the level of single symbols and the level of relations between symbols. For as we have seen, Helmholtz and Hertz claim that objectivity in science is reflected in relations between symbols. This view – the discussion of which will lead us back to structural realism – is also held by the mathematician Poincaré who emphasises the conventional character more strongly than the physicists Helmholtz and Hertz. However, the fact that we *make* symbols rather than somehow *have* them, has already been stressed by Hertz. Indeed this process of representing outside phenomena by symbols is crucial for science, for it allows to establish mathematical or logical relations; i.e. plays an ever increasing rôle especially in modern physics.

Before coming back to quantum field theory, it seems important to say something more about the level of symbolic (mathematical) relations and to introduce the main ideas by a familiar example, namely time. The change of the mathematical relations which hold between symbols is much less radical than the change of the conceptual basis underlying these symbols, especially if it is based on notions of everyday life. In accordance with semiotics, the latter – these rather individual everyday associations – might be called 'connotations'. Whereas the strictly rule-governed inner-mathematical meanings – i.e. the 'mathematical valence', the relationship which the symbol contracts with other symbols of the theory's 'mathematical vocabulary' – would be the 'denotation'.[1] The absolute time in Newtonian physics, for instance, is con-

---

[1] Compare the definition of *denotation* as the 'valence in the semantic field' and of *connotation* as 'emotional or cognitive meaning' in Eco (1968), see also Eco (1975), sect. 2.3.

ceptually very different from that of relativistic mechanics. Our commonsense notions of simultaneity, of the flow of time etc. are in excellent agreement with Newtonian physics but get confused as soon as we start talking about relativistic mechanics. However, things look completely different on the denotative level of the symbols, i.e. on the formal mathematical level where time occurs as a real variable t. Here the kinematical relations of Newtonian mechanics are simply obtained from the relativistic ones by an expansion with respect to the ratio v/c, where v is the velocity connected with a mass m and c the vacuum velocity of light. The relativistic relation (2.1) expressed through the velocity vector v rather than the momentum vector p of a particle is:

$$E = \frac{mc^2}{\sqrt{1 - \frac{v^2}{c^2}}}. \tag{5.1}$$

Performing the above mentioned expansion yields:

$$E = mc^2 + \frac{m}{2}v^2 \left(1 + O\left(\frac{v^2}{c^2}\right)\right). \tag{5.2}$$

The first term is the rest energy, the second term the kinetic energy of a particle in non-relativistic mechanics and the term $O\left(\frac{v^2}{c^2}\right)$ indicates that the relativistic correction is of the order $\left(\frac{v^2}{c^2}\right)$ and thus small if the modulus of the velocity $|v|$ is small compared to the velocity of light c. The corresponding relativistic relation between the time coordinate t' at a point with position vector x' in one (inertial) system $\mathscr{K}'$ and the time and space coordinates t, x in a different (inertial) system $\mathscr{K}$ moving relative to $\mathscr{K}'$ with velocity (vector) v, is given by

$$t' = \frac{1}{\sqrt{1 - \frac{v^2}{c^2}}} \left(t - \frac{v \cdot x}{c^2}\right). \tag{5.3}$$

We see that in the limit where $|v|/c$ tends to zero one obtains $t' = t$, the time becomes 'absolute' in the sense that it does not depend on the inertial system, and hence on the kinematical state of the observer.

We consider it not only remarkable but characteristic, too, that the protagonists of a symbolic interpretation of science, Helmholtz, Hertz, Poincaré and Weyl, are also eminent physicists and mathematicians. For it is very plausible that such a view is first advocated by (mathematical) physicist, who were in a position to experience the permanence in mathematical descriptions; i.e. in the symbolic relations. As opposed to them people judging from outside tend to interpret transitions between theories much more dramatically. Rather than the continuity in mathematical description, i.e. the symbols' denotation, they stress the discontinuities in their connotations. So Thomas S. Kuhn states in the case study of the derivation of non-relativistic physics from the relativistic one that 'those laws have to be reinterpreted in a way that would be impossible until after Einstein's work',[2] whereas in the denotative framework of the theory of symbols the limiting laws, that is the relation between the symbols, *are* indeed the ones of classical mechanics. Thus, the change really concerns only the connotational background of the symbols and not their denotations; i.e. not the 'mathematical valence' of t. Since objectivity in science is reflected not in the symbols themselves but in their relations, Kuhn's arguments on scientific revolutions – or what we would call a 'connotation shift' – might thus be historically correct but do seem epistemologically irrelevant from a physicist's perspective.[3]

As just seen, in some cases it is possible to perform the transition between two theories in a mathematically rigorous way. In that case we can call the limit an effective theory (cf. chapter 2.6). In other cases there are many structural similarities but nevertheless there exists (at least up to now) no derivation in a strict mathematical sense. Coming back to the physical main topic of this essay, the transition from quantum to classical field theory is such a case. We speak of the classical level of a quantum field theory if we make an expansion of amplitudes in powers of $\hbar$, the Planck action quantum, and consider only the leading order. However, the intentions of and questions asked by the two theories are so different that we think it more appropriate to speak of different symbolic representations. The symbols of the two theories are completely different but there are close relations between the structures, e.g. internal or external symmetries. Normally the 'higher' (more fundamental) theory does not make obsolete the 'lower' one. This is obvious for dealing with 'macroscopic' problems. The unsolved difficulties of quantum field theory e.g. do not prevent to solve most complicated problems concerning the propagation of radio waves. It might even be necessary to use the lower theory in the higher one in an essential way, as e.g. in the analytic description of bound states in quantum field theory. In order to calculate the Lamb-shift, one of the triumphs of relativistic quantum field theory, one starts with a quantum field theory in a given 'external' classical field, though this field should in principle also be described by quantum field theory. Thus, to calculate the Lamb-shift one has

---

[2] Kuhn (1970), p. 101.

[3] Similar arguments can be brought forward against Feyerabend's criticism of *consistency conditions*. Feyerabend (1975), sect. 3.

to make use of different symbolic representations at the same time; namely the 'higher' quantum and the 'lower' classical field theory. To elaborate on how and why classical and quantum physics should be viewed as different symbolic representations and why even within quantum field theory there are several symbolic representations, it is helpful to briefly introduce some more semiotic terminology.[4]

5.2 A Semiotic View of Quantum Field Theory

In semiotics a system of symbols with given rules that allow transfer of information is called a 'code'. In this sense not only language is a code, but also, for instance, human perception involves codes (think of the distinction between foreground and background in Gestalt psychology, e.g.); within the arts classicism and romanticism can count as codes and in particular any physical theory put in mathematical terms is a code. For all computational manipulations have to follow certain rules and obviously transport information about external phenomena.

A code has a 'vocabulary' of basic units which – using the code's (syntactic) rules – can be combined to generate larger meaningful combinations. Some codes can be analysed with regard to two structural levels called 'first' and 'second articulation'. The first articulation deals with the smallest meaningful units of the code, the so called 'signs'. In case of spoken language this would be morphemes or words. At the level of the second articulation a code divides into minimal functional units which have purely differential character; i.e. do not have meaning in themselves. They are sometimes called 'figures'. Taking the example of spoken language again, its second articulation consists of phonemes. Thus, the crucial difference between the two levels is that the basic units of the first articulation denote while those of the second articulation lack denotative power.

This semiotic terminology can now be applied to what we think are different symbolic representations in modern physics. Classical mechanics, classical field theory, non-relativistic quantum mechanics and quantum field theory should all be taken to be different codes (indeed within quantum field theory further distinctions have to be made as we will see shortly). All these codes more or less share their second articulation, namely the mathematical language expressed in 'figures' like '$\int$', '$+$', '∂' and further structures. However, they disagree in their first articulation, i.e. in their denoting terms

---

[4] The following terminology is taken from Eco (1968).

and thus also in the entities they describe. Take, as an illustration, electrons, heavy and light quarks. The symbolic systems in which electrons occur include classical mechanics, non-relativistic quantum mechanics and quantum field theory. Heavy quarks only make sense in the context of non-relativistic quantum mechanics and quantum field theory, whereas light quarks only in that of quantum field theory. This is just another (more elegant) phrasing for what has already been stated in our case study on 'Particle and Fields' in chapter 3, namely that questions like 'What is a particle?' or 'Does such-and-such particle exist?' depend on the theoretical background. Given our semiotic approach this simply amounts to say that the denoting terms of different codes might (and indeed sometimes do) vary. Moreover, we think that whether some 'particles' appear closer to our intuition than others depends on the number of codes in which those 'particles' do occur.

We now argue that the various facets of quantum field theory elucidated in chapter 2 do form different codes. The reader is reminded that up to now no quantum field theory describing interaction between particles in four space-time dimensions has been constructed on a sound mathematical footing. Wightman's postulates (axioms) are considered to form the core structure of a theory based on *local* field operators; but they circumvent to formulate the dynamical evolution in terms of field operators. It is this dynamical evolution, however, that entails the *specific physical outcome* of the theory. Nevertheless, the postulates of Wightman allow to derive some stringent structural consequences as e.g. the CPT-theorem, the connection of spin with statistics, and dispersion relations, cf. 2.5.1. These predictions are in fact experimentally very well satisfied and any violation of them would seriously call in question the concept of a local quantum field. Within the Wightman postulates different concrete quantum field theories appear conceivable. Actually, in fictitious worlds having one or two spatial dimensions only, various quantum field theories with internal dynamics have been rigorously constructed and shown to realise the Wightman postulates, cf. section 2.5.3.

A concrete quantum field theory has to be generated by 'quantising' a heuristic classical precursor theory, the Lagrangian density of which is a function of the fields involved and determines the interaction. The Standard Model, in particular, is a non-Abelian gauge theory. As already mentioned above there exists (at present) no fully fledged quantum field theory of this model. Instead, there are two complementary codes to perform a 'quantisation' of the classical precursor: perturbative renormalisation theory, cf. section 2.5.2, and lattice gauge theory, cf. section 2.5.4. The first one, ab initio only formulated as a formal power series based on free quantum fields, cannot create hadrons. Moreover, it employs in addition ghost fields to implement (covariantly) the local gauge symmetry. The second one avoids these unphysical degrees of freedom and starts with a well-defined nonperturbative set-up, which involves only the fundamental degrees of freedom, however restricted to a Euclidean space-time lattice. The respective predictive power of these two codes points to complementary physical domains. The perturbative theory accounts for electroweak processes and to the short-distance behaviour of quantum chromodynamics. In contrast, the lattice gauge theory aims at the long-distance properties of quantum chromodynamics, i.e. the spectrum of hadrons dynamically generated from the fundamental degrees of freedom of the theory.

The respective denotations used in the general theory of quantised fields, in perturbative renormalisation theory, and in lattice gauge theory differ markedly from each other, although they have common elements, too. These differences do not allow to arrange the codes in a hierarchy. The physically distinguished non-Abelian gauge symmetry is not addressed in the Wightman postulates, e.g. In spite of the fact that they could not be realised constructively[5] as yet, they remain a theoretically appealing vision. Both, the perturbative gauge theory and the lattice gauge theory encompass elements of these postulates.

We may look at this state of affairs from a structural realist's perspective. Basing structural realism on the general theory of quantised fields amounts to restrict the theoretical claim to the very core of local quantum field theory. The price to be paid for circumnavigating thereby the formulation of an explicit dynamics is that only few, albeit fundamental and experimentally very well satisfied physical relations result. Besides, the persistent threat to any realistic interpretation by the pessimistic meta-induction may materialise here e.g. if at very small distances (Planck length) the notion of Minkowski space-time becomes obsolete, and hence the Wightman postulates would be subject to revision. A different objection aims at the hidden dynamics in the general theory of quantised fields: as a mere framework it might be compatible with various types of concrete quantum field theories. Or, to express it more emphatically: there might be very different worlds, which all exhibit the structural features resulting from the Wightman postulates. In order to arrive at a wealth of physical relations aiming to cope with the immense body of detailed measurements performed in the subnuclear domain, an adequate con-

---

[5] In four-dimensional space-time.

crete quantum field theory has to be created, i.e. in terms of specific fields and their interaction, thus fixing the dynamical content of the theory. The Standard Model is supposed to play this rôle. As already stated, the 'quantisation' of this model could not yet be achieved fully in the form of a proper mathematical construction. However, partial approaches to this goal are attempted, which appear as facets of quantum field theory with respective codes. To a large extent, the choice between these codes is dictated by the question one has in mind.

This, however, puts a further and much more pressing problem onto structural realism; namely what we might christen the 'problem of synchronic existence'. Remember that structural realism has been introduced to cope with the pessimistic meta-induction. The latter, however, is a problem of 'diachronic existence' of different theories. Structural realism is adapted to explain why the purported entities of different theories which occur one after the other might change; like in the transition from Fresnel to Maxwell, to use the notorious example. However, it is not adapted to explain why there are different theories that exist at one and the same time – as it is the case with the different facets of quantum field theory. For structural realism would have to maintain that one of the facets is the 'real' one. By the same token, structural realism would not be able to cope with the miracle argument any longer. For if it took perturbation theory to be the right structure then the success of lattice gauge theory would become a miracle and vice versa. These approaches, however, can coexist, since they aim at different domains of the full theory.

Indeed it seems that the synchronic existence of different facets is often neglected within the philosophy of quantum field theory and has led to some rather misleading ideas. One is the debate about 'quasi-autonomous domains', brought forward in the context of effective field theories, see section 2.6.[6] The basic idea here is that due to the decoupling theorem – i.e. due to the fact that higher energy scales do have negligible relevance for an investigation at a lower energy scale – the objects of the physical world appear to be layered into separate domains which are more or less independent from each other. At first glance this idea looks like it would provide us with a convenient (stratified) inventory of the world; at one level there are perhaps quarks and electrons, the next level might be inhibited by protons and neutrons etc. However, this is not the case. For the context in which these domains are argued for is purely perturbative. This means that, for instance, strictly speaking not even hadrons

---

[6] These domains have been introduced by Cao and Schweber (1993) and are criticised amongst others by Robinson (1992) and Hartmann (2001).

do occur in any of those domains. Even talking about deep inelastic scattering is of no help since the involved structure functions are nonperturbative entities. Thus, the different facets of quantum field theory have to be taken into account if one was willing to extract a 'full ontology of our world'. However, the idea of quasi-autonomous domains would get really messed up then. For should we assume a separate quasi-autonomous domain for the objects of lattice gauge theory then etc.?

The reason why – at first sight – this notion about quasi-autonomous domains seems to have commonsense on its side can easily be explained by our semiotic approach. For if we have accurate theories on different levels (energy scales), and hence different codes with different denoted entities for each of those levels, it seems quite straightforward to take those entities 'for real' and assume the same hierarchy in their existence as in the theories. However, this step from semiotics to reality (from denotation to existence) is not valid; it is what Whitehead would call a 'fallacy of misplaced concreteness'.[7] First, just because something occurs in a symbolic representation we should not naively take it for 'real'. Second, also a given symbolic system as a whole should not be conflated with reality. The notion about quasi-autonomous domains as derived from perturbation theory violates both. Thus, it seems more promising to stay with a semiotic approach and not to take symbols to refer one-to-one to external objects. This has practical consequences. Consider a sequence of effective theories, pertaining each to a different scale. The cutoff, which is necessary in a non-renormalisable effective theory, would then gain a realistic meaning. This, however, would allow to calculate for instance the energy density of the vacuum, which would turn out to be many, many orders of magnitude larger than the measured one.

In contrast to the structural realism of most of the contemporary protagonists, our semiotic approach – which we developed along the thoughts of Helmholtz, Hertz, Poincaré, and Weyl – does not commit a 'fallacy of misplaced concreteness'. However, coming back to Worrall's initial question[8], alluded to at the beginning of chapter 4, it seems that our approach cannot provide us with 'the best of both worlds'. It abandons a straightforward realist interpretation and thus fails to counter the miracle argument. However, as we have seen, even structural realism fails to account for the miracle argument given that there are different successful facets at one and the same time. A semiotic approach on the other hand is not bothered by such a synchronic ex-

---

[7] Cf. Whitehead (1925), Ch. 3.

[8] Worral (1996).

istence. Additionally, the pessimistic meta-induction poses no problem on it as well. Symbolic systems have always changed and presumably will always be changing.

Admittedly, we did not say much about the process of sign creation and only made some negative claims about their relation to external objects. However, even if we left open some important questions, we have shown the elaboration of a semiotic approach to be necessary for a full account of contemporary physics. The semiotic approach allows to account for the different synchronic facets of quantum field theory; as indeed any serious interpretation of contemporary physics should. Moreover, whereas both structural realism and a semiotic approach run fool of the miracle argument with respect to quantum field theory, the latter at least copes with the pessimistic meta-induction and commits no 'fallacy of misplaced concreteness'.

6

Summary

Examining relativistic quantum field theory we have claimed that its symbolic descriptions of subnuclear phenomena can be understood most adequately from a semiotic point of view. We have emphasised that such an interpretation of a physical theory is not restricted to particle physics, but already came to the fore in the middle of the nineteenth century with the emancipation of the theory of the classical electromagnetic field from mechanics. Notably the semiotic approach goes back to the work of Helmholtz, Hertz and Poincaré, who played a major role in the development of classical field theory. Looking at a theory of quantised fields, the epistemological questions posed with regard to classical field theory are heightened, even if one strictly keeps from the beginning to a probabilistic interpretation of a quantum theory.

Relativistic quantum field theory is widely recognised as forming the adequate theoretical frame to the phenomena of particle physics observed up to the present. Most notably, the 'Standard Model' achieved yields an excellent description of many subnuclear phenomena and made some spectacular predictions later experimentally confirmed. However, a mathematically consistent full-scale construction of a relativistic quantum field theory in four-dimensional space-time, let alone of the Standard Model, is still missing. Therefore, different aspects have to be accessed by particular methods. We have described these approaches as distinct facets of quantum field theory and have subsequently argued to interpret them as different codes which complement each other. As regards the Standard Model, perturbative renormalisation theory accounts for short-distance effects, whereas lattice gauge theory aims at 'large distances', i.e. at those characterising the 'size' of hadrons. It has been pointed out, that the respective relation between quantum fields and associated particles differs within these codes. Moreover, we have considered the concept of an effective field theory and a structural realist's view, which have received particular attention in particle physics and philosophy of science. Evaluating their virtues and deficiencies, we have again stressed the advantages of the semiotic perspective.

It is clear to us that a semiotic approach appears rather general from a realist point of view. We doubt, however, that the various facets of quantum field theory forming the theory of subnuclear phenomena can be subsumed compatible with structural realism. We are certainly aware of general structures visible in quantum field theory as, for instance, the principle of local gauge symmetry. However, this principle provides yet another example of being differently dealt with in the different facets of quantum field theory. In perturbation theory, on the one hand, the concept of spontaneously breaking this symmetry has proven to be of great wealth and led to the discovery of the weak gauge bosons. On the other hand, within the realm of lattice gauge theory, Elitzur has proven that local gauge symmetry cannot be spontaneously broken, unless explicitly by way of fixing a gauge.[1] Thus, once more it is unclear to us, how a structural realist could possibly account for this.

There are strong indications that the semiotic point of view also proves appropriate to assess the analysis of experiments, the host of phenomenological models and theoretical speculations in the realm of particle physics.

The extension of the semiotic approach beyond the scientific enterprise of particle physics, however, is even more fascinating. It is possible in so far as all human and cultural activities can be understood as (transformations of) symbols. Though they differ widely in their respective details, famously such philosophies of symbols have been brought forward by Charles S. Peirce, Ernst Cassirer, Alfred N. Whitehead Nelson Goodman or Susanne Langer.

---

[1] Elitzur (1975).

References

T. Appelquist and J. Carrazone: Infrared singularities and massive fields. *Phys. Rev.*, D11:2856–2861, 1975.

S. Y. Auyang: *How Is Quantum Field Theory Possible?* Oxford University Press, Oxford, 1995.

F. A. Berezin: *The Method of Second Quantization.* Academic Press, New York, 1966.

N. N. Bogoliubov and O. S. Parasiuk: Über die Multiplikation der Kausalfunktionen in der Quantentheorie der Felder. *Acta Math.*, 97:227–266, 1957.

R. Brunetti and K. Fredenhagen: Microlocal Analysis of Interacting Quantum Field Theories: Renormalization on Physical Backgrounds. *Commun. Math. Phys.*, 208:623–661, 2000.

D. Buchholz and R. Haag: The Quest For Understanding in Relativistic Quantum Physics. *J. Math. Phys.*, 41:3674–3697, 2000.

Tin Yu Cao: *Conceptual developments of 20th century field theories.* Cambridge University Press, Cambridge, UK, 1996.

T. Y. Cao: Representation or Construction? An Interpretation of Quantum Field Theory. In T. Y. Cao, editor, *The Proceedings of the Twentieth World Congress of Philosophy*, volume 10 (Philosophy of Science), pages 115–123, Bowling Green, 2001. Philosophy Documentation Center.

T. Y. Cao and S. S. Schweber: The Conceptual Foundations and the Philosophical Aspects of Renormalization Theory. *Synthese*, 97(1):33–108, 1993.

A. Chakravartty: The Structuralist Conception of Objects. Manuscript, 2002.

M. Disertori and V. Rivasseau: Continous Constructive Fermionic Renormalization. *Ann. Henri Poincaré*, 1:1–57, 2000.

H. G. Dosch and V. F. Müller: Renormalization of quantum electrodynamics in an arbitrary strong time independent external field. *Fortschr. Phys.*, 23:661–689, 1975.

Pierre Duhem: *La Théorie Physique.* Vrin, Paris, 1914 (2nd edition, 1957).

F. J. Dyson: The Radiation Theories of Tomonaga, Schwinger, and Feynman. *Phys. Rev.*, 75:486–502, 1949.

54 References

F. J. Dyson: The S Matrix in Quantum Electrodynamics. *Phys. Rev.*, 75:1736–1755, 1949.

U. Eco: *La struttura Absente*. Bompiani, Milano, 1968.

U. Eco: *Trattato di Semiotica Generale*. Bompiani, Milano, 1975.

S. Eidelman and others: Review of Particle Properties. *Phys. Lett. B*, 592:1–1109, 2004.

S. Elitzur: Impossibility of spontaneously breaking local gauge theories. *Phys. Rev.*, D12:3978–3982, 1975.

H. Epstein and V. Glaser: The role of locality in perturbation theory. *Ann. Inst. H. Poincaré*, 19:211–2955, 1973.

L. D. Faddeev and A. A. Slavnov: *Gauge Fields: An Introduction to Quantum Theory*. Addison-Wesley, 2nd edition, 1991.

J. Feldman, J. Magnen, V. Rivasseau, and R. Sénéor: A renormalizable field theory: the massive Gross-Neveu model in two dimensions. *Comm. Math. Phys.*, 103:67–103, 1986.

Paul Feyerabend: *Against Method. Outline of an anarchistic theory of knowledge*. Verso, London, 1975.

R. P. Feynman and M. Gell-Mann: Theory of the Fermi Interaction. *Phys. Rev.*, 109:193, 1958.

K. Gawedzki and A. Kupiainen: Gross-Neveu Model Through Convergent Perturbation Expansion. *Comm. Math. Phys.*, 102:1–30, 1985.

J. Glimm and A. Jaffe: *Quantum Physics*. Springer, New York, 2nd edition, 1987.

D. J. Gross and D. Wilczek: Ultraviolet behavior of non-Abelian gauge theories. *Phys. Rev. Lett.*, 30:1343–1346, 1973.

S. Hartmann: Effective Field Theories, Reductionism and Scientific Explanation. *Studies in History and Philosophy of Modern Physics*, 32(2):267–304, 2001.

Hermann Helmholtz: *Die Lehre von den Tonempfindungen*. Vieweg, Braunschweig, 4th edition, 1877. Engl. translation by A. J. Ellis, Dover 1954.

Hermann von Helmholtz: *Handbuch der physiologischen Optik*. Voss, Hamburg, 2nd edition, 1892.

Hermann von Helmholtz: Die Tatsachen in der Wahrnehmung. In Paul Hertz and Moritz Schlick, editors, *Hermann von Helmholtz, Schriften zur Erkenntnistheorie*, pp. 109ff. Springer, Berlin, 1921.

K. Hepp: *Théorie de la renormalisation*. Springer, Berlin, 1969.

Heinrich Hertz: *Über die Beziehung von Licht und Elektricität*. Strauss, Bonn, 1889. Also in Gesammelte Werke, Bd. I.

Heinrich Hertz: *Die Prinzipien der Mechanik*. Barth, Leipzig, 1st edition, 1894. With a Preface by H. v. Helmholtz.

K. A. Intriligator and N. Seiberg: Lectures on supersymmetric gauge theories and electric-magnetic duality. *Nucl. Phys. Proc. Suppl.*, 45BC:1, 1996. arXiv:hep-th 9509066.

C. Itzykson and J.-B. Zuber: *Quantum Field Theory*. McGraw-Hill, New York, 1st edition, 1980.

F. Jackson: *From Metaphysics to Ethics*. Clarendon Press, Oxford, 1998.

R. Jost: *The General Theory of Quantized Fields*. American Mathematical Society, Providence, RI, 1965.

C. Kim: A Renormalization Group Flow Approach to Decoupling and Irrelevant Operators. *Ann. Phys.*, 243:117–143, 1995.

Thomas S. Kuhn: *The Structure of Scientific Revolutions*. University of Chicago Press, Chicago, 2nd edition, 1970.

J. Ladyman: What is Structural Realism? *Studies in History and Philosophy of Science*, 29(3):409–424, 1998.

G. W. Leibniz: *Mathematische Schriften*. Weidmann, Berlin, 1849.

F. London: Quantenmechanische Bedeutung der Theorie von Weyl. *Z.Phys.*, 42:375, 1927.

J. H. Lowenstein: Convergence Theorems for Renormalized Feynman Integrals with Zero-Mass Propagators. *Comm. Math. Phys.*, 47:53–68, 1976.

I. Montvay and G. Münster: *Quantum Fields on a Lattice*. Cambridge Univ. Press, Cambridge, 1st edition, 1994.

Lochlain O'Raifeartaigh and Norbert Straumann: Early History of Gauge Theories and Kaluza-Klein Theories, with a Glance at Recent Developments. *Rev. Mod. Phys.*, 72:1–23, 2000.

K. Osterwalder and R. Schrader: Axioms for Euclidean Green's functions. I. *Commun. Math. Phys.*, 31:83– 112, 1973.

K. Osterwalder and R. Schrader: Axioms for Euclidean Green's functions. II. *Commun. Math. Phys.*, 42:281– 305, 1975.

Henri Poincaré: *La Valeur de la Science*. Flammarion, Paris, 1905.

Henri Poincaré: *La Science et l'Hypothèse*. Flammarion, Paris, 1927.

H. D. Politzer: Reliable perturbative results for strong interactions? *Phys. Rev. Lett.*, 30:1346–1349, 1973.

S. Psillos: Is Structural Realism Possible? *Philosophy of Science*, 68(Proc.):S13–S24, 2001.

D. Robinson: Renormalization and the Effective Field Theory Programme. In D. Hull, M. Forbes, and K. Okruhlik, editors, *Proceedings of the 1992 Biennial Meeting of the Philosophy of Science Association*, volume 1, pp. 393–403, East Lansing, 1992. Philosophy of Science Association.

Norbert Straumann: Zum Ursprung der Eichtheorien bei Hermann Weyl. *Phys. Bl.*, 43:414, 1987.

R. F. Streater and A. S. Wightman: *PCT, Spin and Statistics, and All That*. Benjamin, Reading, MA, 3rd edition, 1980.

G. 't Hooft: Renormalizable Lagrangians for Massive Yang-Mills Fields. *Nucl. Phys.*, B35:167–188, 1971.

G. 't Hooft: Renormalization of Massless Yang-Mills Fields. *Nucl. Phys.*, B33:173–199, 1971.

G. 't Hooft and M. Veltman: Combinatorics of Gauge Fields. *Nucl. Phys.*, B50:318–353, 1972.

M. Veltman: *Facts and Mysteries in Elementary Particle Physics*. World Scientific, New Jersey, 2003.

S. Weinberg: Phenomenological Lagrangians. *Physica*, A 96:327, 1979.

Hermann Weyl: Was ist Materie? *Die Naturwissenschaften*, 12, 1924. Reprinted Darmstadt 1977.

Hermann Weyl: Electron and Gravitation. I. *Z. Phys.*, 56:330, 1929.

Hermann Weyl: Wissenschaft als symbolische Konstruktion des Menschen. *Eranos-Jahrbuch*, page 375, 1949. Reprinted in Gesammelte Abhandlungen, K. Chandrasekharan, editor. Springer, Berlin, 1968.

H. Weyl: *Symmetry*. Princeton University Press, Princeton (N. J.), 1952.

Hermann Weyl: Über den Symbolismus der Mathematik und mathematischen Physik. *Studium generale*, 6:219, 1953. Reprinted in Gesammelte Abhandlungen, K. Chandrasekharan, editor. Springer, Berlin, 1968.

H. Weyl: *Philosophie der Mathematik und Naturwissenschaft*. Oldenbourg Verlag, Munich, 2000.

A. N. Whitehead: *Science and the Modern World*. Macmillan, New York, 1925.

E. P. Wigner: On unitary representations of the inhomogeneous Lorentz group. *Ann. Math.*, 40:149–204, 1939.

K. G. Wilson: Confinement of quarks. *Phys. Rev.*, D10:2445–2459, 1974.

K. G. Wilson and J. Kogut: The Renormalization Group and the Epsilon-Expansion. *Phys. Rep.*, 12C:75–199, 1974.

J. Worral: Structural Realism: The Best of Both Worlds? In D. Papineau, editor, *The Philosophy of Science*, pages 139–165. Oxford University Press, Oxford, 1996.

E. G. Zahar: Poincaré's Structural Realism and his Logic of Discovery. In J.-L. Greffe, G. Heinzmann, and K. Lorenz, editors, *Henri Poincaré: Science and Philosophy*, pages 45–68. Akademie Verlag, Berlin, 1996.

W. Zimmermann: Local Operator Products and Renormalization in Quantum Field Theory. In *Lectures on Elementary Particles and Quantum Field Theory*, volume 1, Cambridge, MA, 1970. M.I.T Press.

Name Index

Subject Index